Ecological Modernisation and Renewable Energy

Energy, Climate and the Environment Series

Series Editor: **David Elliott**, Emeritus Professor of Technology, Open University, UK

Titles include:

David Elliott (*editor*)
NUCLEAR OR NOT?
Does Nuclear Power Have a Place in a Sustainable Future?

David Elliott (*editor*)
SUSTAINABLE ENERGY
Opportunities and Limitations

Horace Herring and Steve Sorrell (*editors*)
ENERGY EFFICIENCY AND SUSTAINABLE CONSUMPTION
The Rebound Effect

Matti Kojo and Tapio Litmanen (*editors*)
THE RENEWAL OF NUCLEAR POWER IN FINLAND

Antonio Marquina (*editor*)
GLOBAL WARMING AND CLIMATE CHANGE
Prospects and Policies in Asia and Europe

Catherine Mitchell
THE POLITICAL ECONOMY OF SUSTAINABLE ENERGY

Ivan Scrase and Gordon MacKerron (*editors*)
ENERGY FOR THE FUTURE
A New Agenda

Gill Seyfang
SUSTAINABLE CONSUMPTION, COMMUNITY ACTION AND THE NEW ECONOMICS
Seeds of Change

Joseph Szarka
WIND POWER IN EUROPE
Politics, Business and Society

David Toke
ECOLOGICAL MODERNISATION AND RENEWABLE ENERGY

Xu Yi-chong
THE POLITICS OF NUCLEAR ENERGY IN CHINA

Energy, Climate and the Environment
Series Standing Order ISBN 978–0–230–00800–7 (hb) 978–0–230–22150–5 (pb)

You can receive future titles in this series as they are published by placing a standing order. Please contact your bookseller or, in case of difficulty, write to us at the address below with your name and address, the title of the series and the ISBN quoted above.

Customer Services Department, Macmillan Distribution Ltd, Houndmills, Basingstoke, Hampshire RG21 6XS, England

Ecological Modernisation and Renewable Energy

David Toke
Senior Lecturer in Energy Policy,
Department of Political Science and International Studies,
University of Birmingham, UK

First published 2011 by
PALGRAVE MACMILLAN

Palgrave Macmillan in the UK is an imprint of Macmillan Publishers Limited, registered in England, company number 785998, of Houndmills, Basingstoke, Hampshire RG21 6XS.

Palgrave Macmillan in the US is a division of St Martin's Press LLC, 175 Fifth Avenue, New York, NY 10010.

Palgrave Macmillan is the global academic imprint of the above companies and has companies and representatives throughout the world.

Palgrave® and Macmillan® are registered trademarks in the United States, the United Kingdom, Europe and other countries.

ISBN 978-0-230-22426-1 hardback

This book is printed on paper suitable for recycling and made from fully managed and sustained forest sources. Logging, pulping and manufacturing processes are expected to conform to the environmental regulations of the country of origin.

A catalogue record for this book is available from the British Library.

A catalog record for this book is available from the Library of Congress.

10 9 8 7 6 5 4 3 2 1
20 19 18 17 16 15 14 13 12 11

Contents

Abbreviations

ACF	Australian Conservation Foundation
APPA	Asociación de Productores de Energías Renovables
AWEA	American Wind Energy Association
BWE	Bundesverband WindEnergie eV
CAFÉ	Corporate Average Fuel Economy
CCS	Carbon Capture and Storage
CEC	California Energy Commission also Clean Energy Council
CREIA	Chinese Renewable Energy Industry Association
EDF	Environmental Defense Fund also Electricité de France
FERC	Federal Energy Regulatory Commission
FPL	Florida Power and Light
GW	gigawatt
GWe	gigawatt of electricity
IDAE	Instituto para la Diversificación y Ahorro de la Energía
kWh	Kilowatt hour
MRET	Mandatory Renewable Energy Target
MTOE	Million Tonnes of Oil Equivalent
MW	megawatt
NDRC	National Development and Reform Commission
NFFO	Non-Fossil Fuel Obligation
OAT	Office of Appropriate Technology
PPA	Power Purchase Agreement
PTC	Production Tax Credit
PUC	Public Utilities Commission
PURPA	Public Utility Regulatory Policies Act
Pv	photovoltaics
RO	Renewables Obligation
ROC	Renewables Obligation Certificate
RPS	Renewable Portfolio Standard
TREIA	Texas Renewable Energy Industry Association
TWh	terawatt hour

Acknowledgements

I first wish to acknowledge the support of Dave Elliott, who prompted me into drafting a proposal for this book. I would like to thank various people who commented on sections of the book: Neil Carter, Giorel Curran and Richard Cowell, who gave me feedback on Chapter 2; Jan Hamrin and Paul Gipe, who made comments on Chapters 4 and 5 (especially Jan Hamrin, who spent a lot of time on the empirical details). I thank Andrew Blowers and David Humphreys for cultivating my interest in ecological modernisation, Ross Abbinnett for encouraging me to discuss identity issues, and the anonymous reviewers (of the book proposal and the book itself) at Palgrave. I value discussions with Simon Shackley, Dan van der Horst, Xavier Lemaire, Greg Buckman, Miguel Mendonza, Richard Green, Ian Fairlie and also various members of the Claverton Group of Energy Experts – although I stress that this does not necessarily mean that the people I have mentioned agree with what I have said.

Of importance to this work was a series of staff seminar presentations I was given the opportunity to deliver at a range of universities. These include the Science Policy Research Unit at the University of Sussex; the Department of Technology and Social Studies at the University of Maastricht; the Department of Technology Management of Eindhoven University of Technology; the Department of Politics and International Relations at the University of Aberdeen; the School of Geosciences at the University of Edinburgh; the Department of Politics, University of York; the School of Social Policy, Sociology and Social Research, University of Kent; the School of Politics, International Relations and Philosophy, Keele University; the School of Social and Political Sciences, University of Melbourne; and the School of Political Science and International Studies, University of Queensland. I thank the people who agreed to be interviewed. I thank my partner Yvonne Carter for her moral support during this period.

Series Editor's Preface

Concerns about the potential environmental, social and economic impacts of climate change have led to a major international debate over what could and should be done to reduce emissions of greenhouse gases, which are claimed to be the main cause. There is still a scientific debate over the likely scale of climate change, and the complex interactions between human activities and climate systems, but, in the words of no less than the Governor of California, Arnold Schwarzenegger, *"I say the debate is over. We know the science, we see the threat, and the time for action is now."*

Whatever we do now, there will have to be a lot of social and economic adaptation to climate change-preparing for increased flooding and other climate related problems. However, the more fundamental response is to try to reduce or avoid the human activities that are seen as causing climate change. That means, primarily, trying to reduce or eliminate emission of greenhouse gases from the combustion of fossil fuels in vehicles and power stations. Given that around 80 per cent of the energy used in the world at present comes from these sources, this will be a major technological, economic and political undertaking. It will involve reducing demand for energy (via lifestyle choice changes), producing and using whatever energy we still need more efficiently (getting more from less), and supplying the reduced amount of energy from non-fossil sources (basically switching over to renewables and/or nuclear power).

Each of these options opens up a range of social, economic and environmental issues. Industrial society and modern consumer cultures have been based on the ever-expanding use of fossil fuels, so the changes required will inevitably be challenging. Perhaps equally inevitable are disagreements and conflicts over the merits and demerits of the various options and in relation to strategies and policies for pursuing them. These conflicts and associated debates sometimes concern technical issues, but there are usually also underlying political and ideological commitments and agendas which shape, or at least colour, the ostensibly technical debates. In particular, at times, technical assertions can be used to buttress specific policy frameworks in ways which subsequently prove to be flawed

The aim of this series is to provide texts which lay out the technical, environmental and political issues relating to the various proposed policies for responding to climate change. The focus is not primarily on the science of climate change, or on the technological detail, although there will be accounts of the state of the art, to aid assessment of the viability of the various options. However, the main focus is the policy conflicts over which strategy to pursue. The series adopts a critical approach and attempts to identify flaws in emerging policies, propositions and assertions. In particular, it seeks to illuminate counter-intuitive assessments, conclusions and new perspectives. The aim is not simply to map the debates, but to explore their structure, their underlying assumptions and their limitations. Texts are incisive and authoritative sources of critical analysis and commentary, indicating clearly the divergent views that have emerged and also identifying the shortcomings of these views.

The present volume is very much in the latter category. It attempts to both challenge and develop the theory of ecological modernisation that has emerged in recent years. Ecological modernisation has often so far been seen as a prescription for making relatively minor adjustments to how energy policy was developed and energy projects carried out, often in a 'top-down' way. Toke, however, suggests that, not only has the modernisation that has occurred, for example, in relation to the adoption of wind power, involved in its earlier development the essential element of social movements, but that the comprehensive industrialisation of renewable technologies involves the development of a separate 'eco identity' as well. Toke says that this identity, with which the public and environmental NGOs sympathise, means that, in reality, ecological modernisation (EM) proceeds differently to that of conventional ecological modernisation theory. In conventional EM, mainstream industry (such as the electricity industry) makes the key technological choices. In Toke's notion of 'identity EM', technological choices are prompted by 'bottom-up' pressures for specific technologies with a sustainable energy identity. This distinct technological identity persists even now that renewables are becoming major actors in the world energy industries. Moreover the renewable industries act as competitors with conventional fossil fuel and nuclear energy sources seeking to replace them in part or whole. Toke sees this 'identity EM' process as central to the future continued success of renewable energy technologies.

Dave Elliott

1
Introduction

The primary aim of this book is to develop political analysis of environmental policy issues using the case of renewable energy. The two central questions are: a) how have new renewable energy sources, led by wind power, developed and expanded in recent decades? b) how can we deploy, and also modify, ecological modernisation theory to help answer this previous question?

Although the prime focus of the development of theory is ecological modernisation (EM) theory, other modes of political analysis also need to be discussed in order to supplement the analysis. This theoretical development is important if we are to grapple with one of the key sustainability issues of our time, that is, how renewable energy has been developed through different policy frameworks and contexts, and what and how different outcomes have occurred in different states and countries.

There will be an extended discussion of 'ecological modernisation' in Chapter 2. However, essentially, the idea of EM is that it combines economic development and environmental protection as a way of conducting good business: 'in short, business can profit by protecting the environment' (Carter 2001, 226). EM's central theme is that consumers demand higher-quality, environmentally sustainable goods and services and business responds to this pressure, so increasing economic development. This produces a 'positive sum' solution whereby economic development is increased rather than decreased by environmental protection, and where policy 'conceptualises environmental pollution as a matter of inefficiency, while operating within the boundaries of cost-effectiveness and administrative efficiency' (Hajer 1995, 33).

EM demands a holistic response by industry to environmental problems; that is, policies must be considered for their total environmental impact, rather than environmental policy being limited to one-off 'end of pipe' responses (Weale 1992). EM can be said to be a theory that underpins a lot of the 'precautionary' legislation that businesses need to implement, and in doing so business is regulated by bodies such as the Environmental Protection Agency and the European Union through its directives. As will be discussed, existing interpretations of EM are unsuitable for the renewable energy case study, and my purpose in this regard is to revitalise EM by introducing a new notion of 'identity EM'.

Deploying an EM framework can help us understand our policy options for the future. As we shall see, the 'modern' spur for renewable energy came from the oil crises of the 1970s. This book is written in the aftermath of a further oil crisis (2008), one that may be succeeded by another oil crisis before the global energy economy is restructured away from depleting conventional oil supplies. Just as in the 1970s, we face choices about paths, sharpened in today's world by climate change considerations as a policy driver. Theoretical frameworks may help understand how we can navigate the politics of these paths. Ecological modernisation occupies a pivotal role in ecological transformation, although 'the jury is still out on whether or not ecological modernisation offers a practical programme for achieving sustainability' (Carter 2001, 222). This book contributes to this debate through a reappraisal of the theory of EM in the context of the case study of renewable energy.

I argue, in the case of renewable energy, that it is public support for renewable energy as a new set of technologies that is the key political driver for its development. In another work (Toke 2011) I criticise mainstream EM for a failure to place sufficient emphasis on the involvement of social movements in the development and deployment of renewable energy. In this work EM theory is developed into what I call 'identity' ecological modernisation. This includes the notion that it is positive public identification with specific (renewable energy) technologies that has allowed the emergence and growth of the renewable energy industries as an alternative sector to that of conventional energy industries such as fossil fuels and nuclear power. This new set of (renewable energy) industries is often in competition with major energy corporations, and certainly in competition with 'conventional' fossil fuel and nuclear power industries.

This phenomenon of public identification with an industry with perceived ecological benefits has more significance than just being an advantage for renewables – public support is essential for renewable development, since without the public support the incentives needed for the current scale of development would not materialise. Hence, if EM is to be used as an analytical framework to discuss renewable energy development, EM has to have a clear means of conceptualising and analysing technological identity. This argument is articulated, in Chapter 2, in the context of a critique of major works in the existing literature on EM.

A means of measuring the extent of 'identity EM' in different country-case studies is drawn up and a relevant use of discourse, interest group and institutional analytical tools is discussed. The theoretical approach is then applied to several case studies.

An outline of later chapters

As I have already mentioned, Chapter 2 deals with the central theoretical topic of this book, ecological modernisation (EM), and also other relevant analytical tools. Chapter 3 charts the genesis of renewable energy technologies, focusing mainly, although not exclusively, on the most implemented renewable technology so far, wind power. The nature of renewable technological identity is discussed and changes in perceptions of identity are studied. How wind power in particular was transformed from being a failed technology into a 'new' energy source, and the associated emergence of this new energy identity in the aftermath of the 1973 oil crisis, is carefully analysed. This transition is closely associated with being an alternative to nuclear power. Wind power in its modern form grew up in Denmark, and its emergence here is explained by reference to Danish traditions of rural self-sufficiency and community cooperation and the mobilisation of 'non-material' influences.

Since the 1970s and 1980s renewable energy has been accepted into the mainstream energy agenda in an increasing number of countries, and hence its identity has changed from being so much associated with the anti-nuclear movement. Even ecologists' support for renewables has shifted in that, for example, the notion of building interconnectors to shift variable energy supplies across continents is being accepted. However, there is still argument about how far renewables can be absorbed into the conventional energy system without fundamentally changing that system.

Arguments about renewable energy at a local level can be conceptualised as being about conflicts between different identities, for example between public identification with wind farms as clean energy sources and, alternatively, as being in conflict with idealised notions of local place identity. Deliberation is found to have limits in acting as a solution to such identity conflicts.

Chapter 4 concentrates on the development of the first big industrial market for wind power in California. There is a discussion of how an anti-nuclear movement emerged in the aftermath of a vigorous anti-nuclear movement which itself emerged in the context of a strong environmental tradition. The election to Governor of a liberal Democrat, anxious to please liberal activists, led to administrative changes which induced the utilities to give contracts to independent renewable energy companies. This allowed the small wind industry that had emerged in Denmark to grow with a much bigger market. The characteristics of identity EM are much in evidence in the Californian drive for renewable energy in the 1970s and 1980s.

Although the 'wind rush' petered out in the late 1980s, it left a legacy in the form of an example of a new renewable presence. The main elements of 'identity' EM can be seen in the case of California, including the influence of 'non-material' factors. However, since the 1980s the growth in renewable energy has, until very recently, been disappointing. In the 1990s attention shifted towards seeing the introduction of competition in energy provision as the priority.

Chapter 5 examines the factors associated with the development of renewable energy in states with the largest installed wind power capacities. Two key ingredients for success are identified: the Federal-based production tax credit system and RPS measures. While the RPS systems have been billed as 'market-based' instruments, their main utility has been in providing a means of altering the regulatory structure so that the utilities feel they have to invest in renewables or fail to achieve other regulatory objectives which they prefer. Trading in renewable electricity certificates makes little or no difference to the practical operation of RPS systems that are studied in Texas, California, Iowa and Minnesota. California has a highly complex RPS, which is likely to fail to live up to its targets, as the utilities appear to attach different tests to whether to give contracts for wind power projects compared with solar power. By contrast, Texas has a much more rapidly expanding demand for electricity and much fewer administrative constraints on wind power expansion. However, there are few incentives for renewables other than wind power. There is clear evidence of several of the characteristics of 'identity EM' in the cases of the four states covered. These include dedicated incentive structures for renewable energy, independent renewable lobbies, coalitions between renewable trade lobbies and NGOs opposing utility interests and, to a varying extent, development being done by independent companies.

Chapter 6 analyses five further countries, again using 'identity EM' as a framework. These are Spain, Germany, the UK, Australia and China. These countries are selected in order to provide a contrast between countries that all lay claim to having significant renewable energy programmes. Spain and Germany are acknowledged to have advanced renewable energy programmes; the UK programme has taken longer to accelerate, Australia still longer and China is emerging as a major world player in the renewables industries. Germany's renewable expansion, like that of Spain, was begun in the context of strong anti-nuclear sentiments. An important reason for the inclusion of China is not just the importance of that country, but also because it presents us with an opportunity to deploy the 'identity EM' framework in a non-Western, still developing, country. In some respects the results of the analysis are surprising, in others perhaps less so. China it seems, fulfils some of the characteristics of 'identity EM' in ways that are comparable to other cases and to a greater extent than is the case with Australia.

Indeed, there are, in some ways, greater similarities between, say, the cases of Spain and China than between Spain and Australia, including rapidly increasing energy demand and demand for 'clean energy' sources. Key institutional factors that underpin the different outcomes involve perceptions of energy security and the strength of social movements for green energy. A social movement has promoted renewable energy in Germany, and in the 1980s in Spain, although energy security drives a consensus for renewable energy in Spain today, and also, increasingly, in the UK. Australia has less of an impulse to act on energy security concerns because of its perceived identity as a coal exporter.

The concluding Chapter 7 reflects on the results of the use of the theoretical tools, first 'identity EM', and then institutionalism and traditions. The existence of characteristics of 'identity EM' in particular country-case studies is found to be highly associated with the extent of the renewable energy programmes. The debate about 'market-based' or 'command and control' policy instruments is dismissed as a diversion, to the extent to which the discussion needs to be about the most suitable institutions for the promotion of renewable energy. 'Identity EM' is associated with feed-in tariffs as a policy instrument to promote renewable energy, since it involves public identification with, and support for, specific renewable energy technologies.

In discussing these issues I want, in this volume, to focus on the interpretative side of the debates about the emergence, role and importance of renewable energy. I include numbers where appropriate to explain understandings or how positions are justified. However, there is a surfeit of number-crunching technical analyses of renewable energy. Many of them are highly enlightening and important, such as that produced by Boyle (2004). I have no need to repeat this type of exercise. Renewable energy has not emerged and grown in its modern industrial form simply because technicians have crunched numbers, as important as this sometimes may be to justify value positions. Indeed, the conventional energy industry, which was originally rather indifferent to renewable energy, has always claimed a monopoly of technical wisdom. So how is it that renewable energy has grown? The answer is a political one, and one that requires analysis of history.

Conclusion – research questions

We need to unpack the basic theoretical and empirical aims of the book that were stated at the start of this chapter, perhaps by pointing out a few bullet points. Research questions, therefore, include:

- How suitable are existing versions of ecological modernisation (EM) theory in providing a framework for understanding the renewable energy case? What problems arise when existing versions of EM are applied to the renewable energy case?
- What is the nature of the 'identity' EM that is proposed to ameliorate these problems with EM; what are some characteristics of identity EM that can be used to measure the extent of identity EM in case studies?
- What other tools of analysis can be applied to this renewable energy case study?
- How has the 'identity' of renewable energy been constructed and how has it changed?
- In the light of this, in what circumstances has renewable energy developed most effectively? What lessons might this have for the future?

2
Revising Ecological Modernisation Theory

At the end of March 2009 the UK's Energy Minister, Mike O'Brien, opened a Conference on the possibilities for UK Ports businesses offered by the prospect of 25 per cent of the UK's electricity supply being met by the Government's programme of offshore wind power. He said:

> We need to bring about a revolution in the way energy is produced...Imagine you are pin-striped revolutionaries in the spirit of Che Guevara on the Sierra Madre. (O'Brien 2009)

Less than a month later, the Chairman of the Federal Energy Regulatory Commission declared that the USA could in future substitute renewable energy for nuclear and coal-fired sources of electricity. 'We may not need any [new nuclear or coal plants] ever,' said the FERC Chairman Jon Wellinghoff (Johnson 2009).

All of this may be dismissed as (partly) public relations bravado, but the contrast with the position of the nascent wind power and new renewable energy movement thirty years previously is striking in the sense that times have moved on. Then the grass roots were calling for an energy revolution from below, but now renewable energy is being promoted by policies from above. As we shall see in later chapters, many anti-nuclear radicals of the 1970s saw renewable energy as a decentralised alternative to the centrally inspired notions of large power stations. Yet here are government ministers and CEOs of key energy regulators trying to inspire representatives of corporate capitalism to invest in technologies, particularly wind power, that have now become very large-scale. Moreover, they are inciting such activity in pursuit of a Government strategy that has an avowedly environmental objective, that of tackling global warming as one of its supreme objectives. This chapter aims to answer the question: to what extent can existing, or possibly adapted, versions of ecological modernisation (EM) help us understand the emergence and development of renewable energy as a policy and political phenomenon?

As was suggested in the Introductory Chapter, EM may be succinctly defined as one involving an approach, by business, to environmental protection through technological development, which improves both economic development and environmental protection to be a sum than is greater than its parts. It has come to encapsulate EU strategy (Murphy and Gouldson 2000). Toke (2002a, 147) lists six points as being central to EM:

> First is the idea that economic and environmental objectives can be simultaneously achieved with a positive sum outcome. Second, economic development and ecological protection are both desirable objectives for the welfare of both present and future generations. Third, is the 'polluter pays' principle. Fourth is the notion of a 'holistic' approach to problem-solving that dismisses any idea that environmental problems can be dealt with individually. Fifth, environmental protection policies need to be dealt with in a market context, but accompanied by government intervention. Finally, nations need to adopt ecologically sustainable policies in order for them to compete effectively with other countries.

As Barry and Paterson (2004) have pointed out, the EM discourse has proved to be attractive to avowedly 'modernising' governments such as New Labour. Following on from Mol (2001), they argue that globalisation tendencies have both enabled and limited the possibilities for New Labour's environmental reforms.

EM may apply most to richer industrial economies, but it may also apply to an increasing extent to developing economies that are attempting, through market pressures and also governmental design, to incorporate environmental objectives into their development paths.

Ecological modernisation is clearly distinguished from so-called 'dark green' approaches by having a positive attitude to economic development, provided it can combine this with an ecological rationality. Ecological modernisation also has a positive view of the use of technology to deal with environmental problems, although that is far from saying that this is the only significant aspect of ecological modernisation (Mol 2006). However, in this it is sharply distinguished from dark green views. According to Dobson (1995, 96): 'What can be said, it seems to me, is that wholehearted acceptance of any form of technology disqualifies one from membership of the dark-green canon.' However, as we shall see, taking the side of ecological modernisation in this dispute leaves various questions of emphasis unanswered: in particular, who defines the relative priorities given to ecological and developmental objectives, and, connected with this, which technologies are chosen and who exercises the technological choices.

The importance of finding answers to this question lies both in helping to understand renewable energy policy outcomes and also in improving understanding of ecological modernisation theory. In doing so, we can reflect on the utility of different versions of EM. As will be discussed, I focus on analysis on two existing variants of EM in particular – a so-called 'mainstream' version and a so-called radical version – and I outline a third notion of 'identity' EM, which may be used as a template for analysis of the development of the renewable energy industry.

EM theory is used here in the form of a way of analysing how environmental change is (or is not) incorporated in business and the economy. According to Carter (2001, 213): 'Perhaps the most distinctive feature of ecological modernisation is that it directly addresses the business sector, whose support...is vital for any transition towards a more sustainable society.' It is to be distinguished, at least in what is here called the 'mainstream' form, from sustainable development by its Western focus and its relative lack of focus on equity issues (Langhelle 2000).

I attempt to set out an account of how ecological modernisation is played out in the case of renewable energy. In doing so I relate my approach to what Buttell (2000) calls a) the 'objectivist' and b) the 'social constructionist' approaches to EM. He identifies the mainstream variant of EM as being identified with authors such as Huber (1991; 2004), Janicke (2008), Mol (1995; 1996) and Mol and Spaargaren (2000). Buttell describes this approach as 'objectivist' in the sense that EM is used to analyse how ecological change occurs as, most basically, a process whereby conventional industry adapts its technologies and practices as a response to social pressures to achieve environmental objectives. The 'social constructionist' (or what I call 'radical') approach, on the other hand, analyses EM as discourses about environmental policy that are objects of critique and can only result in sustainable outcomes if 'bottom-up' patterns of deliberation are involved. The discourse–deliberation approach is articulated by Hajer (1995). Christoff (1996) introduces a distinction between 'strong' and 'weak' ecological modernisation. This normative approach in many ways extends and clarifies points made by Hajer (1995) and is thus also labeled as being in the 'social constructionist' camp of EM. This social constructionist EM approach is 'radical' in the sense that it challenges the more industrially conventional 'mainstream' version.

Mainstream EM

My own use of EM in analysis of renewable energy development can be expressed after performing a critique on both of these (mainstream and radical) approaches to EM. First I want to talk about mainstream EM as typified by Mol's (1995) contribution. This work on the Dutch chemical industry (1995) represents a formidable argumentative and empirical illustration of EM. It is for that reason that I use it as a key focus for critique of the mainstream EM position. His argument is a clear one, although the extent to which its various claims are justified is a different issue.

Mol reviews the development of EM theory, starting essentially with Huber (1991), and outlines a sociological context and underpinning for his approach. He contrasts an ecological modernisation approach with 'neo-marxist' and 'de-modernising' approaches. He then sets out some hypotheses for discussion and a methodology, and he then uses that methodology to analyse the chemical industry through its main sections: paint, plastics and pesticides. In doing so Mol discusses six hypotheses (1995, 58), summarised as: production and consumption are increasingly subject to 'ecological criteria' as well as 'economic criteria'; science and technology increasingly penetrate industrial systems in achieving such criteria; private market processes and 'negotiated rule-making' supersede top-down bureaucratic rule-making; environmental NGOs change their ideological approaches from opposition towards being more engaged in direct negotiations with industry to achieve ecological reform; 'ecological restructuring' is spread through globalisation; and 'alternative de-industrialisation initiatives' are thinly supported and rarely applied. Mol utilises conventional interest group and policy network analysis to underpin his approach.

His approach focuses on how an existing, established industry responds to ecological pressures. Within, for example, the paint industry, the trade organisation representing the Dutch paint industry, the VVVF (Mol 147–53), is representative of the paint industry as a whole and is lobbied by environmental interests, which latterly are more interested in elite discussions on policy details than in outsider campaigns of opposition. The VVVF plays a leading role in achieving the diffusion of techniques that ameliorate environmental problems.

There are companies that are more farsighted in environmental terms than others, for example, and that might use processes and products that result in emissions of fewer PCBs or lower quantities of volatile organic compounds (VOCs). However, a feature of EM is that progressively adapted legislation tends to ensure that these processes become standard practice in the industry. Soon 'good' practices become diffused as a result. The central point here is that change is engineered by an existing conventional industry, which chooses the techniques necessary to achieve the environmental standards contained in new regulations and agreements.

While I have sympathy with many of the elements of what Mol calls his 'sensitising hypotheses' (Mol 1995, 58), I have problems with some of what is there, what is omitted and also the analytical status of the theory. The first problem is with the claimed objectives and methods of Mol's approach to EM.

Objectives and methods

The objectives of the 'mainstream' EM theory are unclear (Mol 1995, 436–8). The objectives are sometimes conflicting, the framework may not be capable of delivering all of them, and the methods will not deliver all of the objectives. At one point Mol says that 'Ecological modernization theory claims to be capable of…predicting how and why environmental reform takes place.' He then says that social theories (including EM) are evaluated through 'argumentative discourse', but later on he states that 'the pre-interpreted subject matter of the EM theory (i.e. the environment-induced transformations in social conduct and institutional forms) may interfere with both its predictions and explanatory understanding as summarized into the six "hypotheses". Consequently these hypotheses should not be interpreted as statements to be falsified on the basis of empirical evidence, but rather as sensitizing theorems which theoretically inform the complex social world …' (Mol 1995, 436–8).

So, according to this, EM theory can predict outcomes but we cannot test these claims. Is not this a bit like saying 'we are right in what we say, but there is no way you can prove us wrong because it is a matter of interpretation'?

While Mol may do a masterful job of illustrating his hypotheses in the case of the Dutch chemical industry, his framework is wanting if it claims to be transferable to all other cases of industrial adaptation to ecological pressures. I argue below that Mol's approach to the chemical

industry is inappropriate as a means of analysing the renewables industry. Of course, a defence of the mainstream EM position may be to argue
that the EM model can be used differently in the electricity and renewables case to take account of points that I make. Yet, but if this is the
case, and if the mainstream EM model can be interpreted differently
so as to fit a different case study, then the claims to provide objective,
generalisable analysis and predictions are rendered meaningless. Mol's
study (1995), and the hypotheses it contains, become applicable only
to the case study upon which he focused, namely the Dutch chemical
industry in the period he studied.

Mol asserts that the analytical and normative claims of EM can be
discussed separately, but if the analytical claims of EM cannot be adequately assessed, and we are not even sure how EM does describe what
is going on because the theory can be interpreted in different ways,
this does not bode well for a 'normative' project of discussing whether
it should be the right strategy for achieving ecological reform. We cannot have a debate about whether something 'should' happen if we do
not know what the 'something' is! Of course, we can also assess the
relevance of Mol's (1995) model by applying it to another case study
and attempting to transfer the claims made for the Dutch chemicals
industry to the renewables case study. I shall do so.

Structure of EM models

However, perhaps the biggest problem with trying to apply Mol's framework to the renewable energy case is that the structure of his model is
inappropriate for use in the case of renewable energy. By Mol's 'model'
I mean the political agents/actors he selects, the role these agents/actors
perform, the various relationships that are said to exist between the
actors, and the processes in which these actors/agents are said to engage.
Problems emerge if we try to transfer the same type of agents, relationships, processes and roles of agents used in the Dutch chemical industry
case to the case of renewables. This becomes clearer though a discussion about the model which Mol implicitly uses in his analysis. The
model that is selected can have a significant impact on the conclusions
of the analysis. In Mol's study (1995) the central agents are the chemical
industries (divided into paint, plastics and pesticides), the trade associations representing these sectors, and environmental NGOs. The central
process described is one of the three chemical industries accommodating, on a generally incremental basis, demands for reducing the environmental impacts of their industrial activities.

I argue that it is necessary to use a different model in the case of renewable energy, in the sense of talking about a different set of industrial agents, processes and relationships compared with the chemical industry case. Whereas in the chemical industry the conventional industry is the central type of agent, in the case of renewable energy described in this book there are two types of central agent. One, for sure, is the conventional electricity industry (divided into different fossil fuel, nuclear, old hydro generation and the electricity distribution and retail sectors), but another crucial central agent can be identified as the renewable industries themselves, wind power, solar, photovoltaics (pv) and so on. The new renewable industries are emergent ones, and are agents that have a distinctly different role with regard to ecological reform, given that a key part of such reform is conceived as the deployment of renewable energy. It could, therefore, be argued (and indeed is argued in this book) that the renewable energy industries need to be analysed as distinct entities compared with the conventional electricity industry. In addition, their emergence needs to be explained. Environmental groups can be seen often to be closely allied to renewable energy trade groups for the purposes of persuading the state and the electricity industry to give incentives to promote renewable deployment.

Grass roots activism was necessary to develop the beginnings of an industry that the major industrial players were otherwise ignoring. The renewables industry had to struggle against the electricity establishment in the early years, and even now it has to compete with what is still a dominant fossil fuel and nuclear industry for resources, usually gaining those resources through governmental legislation (often) backed by public campaigns. Environmental groups may today act in support of the renewable trade associations rather than taking the lead themselves, but here it is clear that they work more closely with the renewables trade associations than those of the conventional electricity industry.

Yet this is rather different from the model associated with Mol's analysis of the chemical industry. Applied to the renewable case, there would be a focus on the 'electricity industry' as a central unit that develops schemes for reducing environmental impact through processes of 'modern science and technology' (Mol 1996, 313), and that the renewable technology trade groups should be treated as part of the conventional electricity industry. In this alternative model, which may be implied by Mol's chemical industry model, another type of agent is typified as grass roots idealistic activists whose attempts to bypass conventional industry are doomed to failure and who are superseded by more professional groups who have measured negotiations with the electricity industry about environmental reform.

This type of approach is implied by various pointers in mainstream EM accounts. For instance, there is a thinly veiled dismissal, by Huber, quoted by Mol, of the grass roots ecological technological initiatives (of the 1970s), which were said to have involved a 'deception...that although we were remarkably successful in various fields, we never matched the initial goals' and that 'The alternative enterprises were never competitive and always lived on the resources of other people,' that capital was in short supply, the machinery was inefficient and expensive and 'producing sub-standard products no-one on a regular market wanted to have' (Huber 1991, 182, cited by Mol 1995, 35–6).

Hence one arrives at this type of description:

> The initial close connection between the 'soft' character of technologies on the one hand and their small scale application in the context of a decentralized organization of production on the other, seems to be broken up now that for example windmills and solar energy are incorporated into the environmental policies of electricity companies...When soft-path technologies are propagated outside the market their chances of survival seem to be less in comparison to the situations in which they are incorporated in the strategies of major industrial actors. (Spaargaren 2000, 52)

This type of description carries some potential normative interpretations: that grass roots technological forays are best left to the 'conventional' industry of 'major industrial actors', and, moreover, that the term 'electricity industry' can be used to lump together the interests of both the renewables sector and the rest of the electricity industry. Yet this interpretation is contested by the narrative contained within this book as being at best being simplistic and often simply wrong.

Three further points can be made in criticism of Spargaaren's quote. First, the mainstream EM narrative may imply an understatement of the early importance of the need to develop the industry from small beginnings with idealistic support as a necessary precursor to industrial transformation – as opposed to any implication that this would have been better done by the conventional energy sector from the start. Second, there may be misunderstanding of the early beginnings of wind power in particular, in Denmark, which was implemented as a grid-connected technology, that is, connected to mainstream industry, not merely as a means of energy self-sufficiency. As will be discussed in Chapter 3, the meaning of the term 'decentralisation' has a wide number of interpretations, and this does give it power as a mobilising discourse for various interests, including commercial organisations, who developed a wider industry. Thirdly, as discussed earlier (and expanded upon in later chapters), even in later stages of commercial development, renewable energy lobbies often need to be discussed as discrete industries trying to transform the electricity industry. This is as opposed to analysing change as a process whereby an electricity industry, viewed as mainly being a homogeneous bloc, is persuaded by external environmental pressures to adopt the renewable technologies. There are strategic divisions within the industry, and this needs to be reflected in the mode of analysis and a crucial structural distinction made (in the model used) between the renewables lobby, as representing distinctive interests, and the conventional electricity industry.

Moreover, there is some evidence even within Mol's (1995) study that even when there is evidence of the emergence of an alternative technological paradigm there is a tendency to pay little attention to this alternative. Mol discusses the pesticides industry as one of three empirical sections that make up his treatise, but spends barely half a page discussing the organic industry. He comments that organic agriculture 'has become the dominant paradigm of the major part of the Dutch environmental movement...Its relative success challenges the dominant agriculture, forcing it to restructure itself towards an integrated pest management agriculture' (Mol 1995, 334). That may be the case, but the point is that Mol spends almost the entirety of his study analysing a conventional industry rather than a technology (organic agriculture) whose practitioners want to replace, to a greater or lesser extent, the conventional chemical-intensive agricultural paradigm, not merely reduce the amount of pesticides used in conventional agriculture. The relationships between the environmental NGOs and the organic alternative are rather different from those in the case of environmental NGOs and the conventional chemicals industries, but this is accorded little emphasis or significance in Mol's general conclusions. What matters in this case is not just the substance of the empirical analysis, but the structural makeup of the model used to analyse the case studies. This will influence the analytical conclusions of EM narratives and implicitly affect normative judgements, since the analysis alters the agenda of choices upon which the normative decisions are made.

There is a plausible case to argue that Mol's EM 'sensitising hypotheses' do not pay sufficient attention to the emergence of alternative 'ecological' technological paradigms, and this in turn will influence the normative agenda, for instance concerning whether technological ini-tiatives should be encouraged outside the realms of conventional indus-tries. At least one of the hypotheses may be open for direct criticism in the light of the arguments contained within this book. This concerns the relationship between industry and environmental groups.

Industry and environmental NGOs

The fourth of Mol's six hypotheses for the working of EM theory is:

> environmental NGOs change their ideology, and expand their tradi-tional strategy of keeping the environment on the public and politi-cal agendas towards direct participation in direct negotiations with economic agents and state representatives close to the centre of the decision-making process, and the development of concrete proposals for environmental reform. (Mol 1995, 58)

This may serve well for the purposes of Mol's analysis of the chemi-cal industry, but it may not help us so much in understanding how environmental groups and civil groups are involved in renewable energy. As we shall see in Chapter 3, the involvement of the envi-ronmental groups, especially anti-nuclear activists motivated to involve themselves in renewable energy activism, was especially intense in the early days of modern renewable energy, and this activity manifested itself closely in making decisions about the technologies themselves rather than relying on the energy indus-try to do this. This occurred, as we shall see, in different ways in places such as Denmark, California, Spain and Germany during the 1970s and 1980s. Environmental groups may or may not have changed their ideology on renewable energy (even this is debata-ble given the continued emphasis on 'decentralisation'), but given the discussion in later chapters it is difficult to maintain that they have become 'more' concerned with 'concrete proposals for envi-ronmental reform' and direct negotiations with the conventional electricity industry – groups such as the Friends of the Earth and the Environmental Defense Fund have been consistently concerned with such issues since the 1970s. These days pro-renewable NGOs oper-ate in a coalition with the renewable energy trade groups, often in public campaigns to pressure the conventional industry into accept-ing greater incentives for renewables. Even now, grass roots activists are involved in various ways directly in organising the technology itself, through widespread ownership of wind farms in Germany, through citizens acting as user–generators themselves when solar pv is installed on rooftops.

As we shall see in later chapters, even today, when many renewable development companies are owned by divisions of the energy multinationals, these renewable companies still operate through trade associations that push for renewable energy interests, often in opposition to fossil and nuclear interests and pressures by electricity suppliers, for example to keep electricity prices down. As will also be discussed in later chapters, renewable subsidiaries have their own identities and interests, which are sometimes noticeably in conflict with their parent electricity companies. These renewable companies and trade associations ally themselves with environmental and citizen NGOs to support giving incentives to renewable energy.

A central feature of the battles in support of wind power, solar power, farm-based biogas and so on is precisely that they are usually economically unviable without the financial support mechanisms that are deployed to support them. Environmental NGOs and other public advocates mount public campaigns for these incentives rather than relying on the energy industry to implement or push for such incentive structures entirely of its own volition.

That leads on to another big difference in comparison with, say, analysis of the chemical industry. Chemical companies will be subject to the same environmental legislation and standards, but renewable energy is supported by a specific set of incentives and institutions crafted to promote its interests rather than the other fuel sources. Lobbying tends to be more or less constantly focused on such issues, and it is here that a specific set of coalitions are formed between renewable interests and environmental groups, and sometimes wider alliances of citizen and social groups. At best the mainstream EM analysis is simplistic if applied to the renewables sector. Environmental NGOs do not negotiate with the electricity industry so much as support an emergent part of the energy industry, renewables, against the claims of the conventional industry.

Economic and ecological rationalities

In their strictures about the need for doing projects according to 'economic viability', authors in the 'mainstream' EM school may sail into controversial normative waters. According to Mol (1995, 33), 'The ecological modernization theory does not perceive of the emancipation of ecological rationality or the ecological sphere over or instead of the economic rationality and sphere...The social practices of production and consumption should be designed and evaluated according to – at least – both rationalities.'

This raises the question of how this is interpreted. Are the renewable technologies to be assessed against the economics of conventional power sources? Are they to be assessed against other renewable sources? Many would argue that these energy sources should be developed because they are the right technologies from an 'ecologically rational' point of view, regardless of their cost. Costing 'ecological rationality' for economic viability is fraught with problems, especially given the wide disagreement about estimates for external costs of energy sources. In practice economic rationalities are important – for example, in Spain in 2009 the solar pv programme was pruned in the context of a recession and concerns about rising electricity prices – but decisions about what incentives and what institutional constraints are given to different technologies are inevitably bound up with value judgements, and they are not, and, from an ecological point of view, should not be, dominated by considerations of current market cost. It is not possible, in a world of uncertainties about future costs of energy technologies and future policy priorities, to declare with certainty what will be viewed as 'rational' in fifty years' time. Given this uncertainty, an ecological programme has plenty of room to make the ecological rationality take precedence over the economic rationality, which is itself full of uncertainties.

If emphasis is given to an economic rationality independent of an ecological rationality, then extra incentives may be given solely to the renewables options that are cheapest at the moment, or, perhaps, not even to renewables that are relatively cheap but need long-term guarantees of income streams to justify capital-intensive investments. For example, currently, solar pv is an expensive technology to support relative even to some other renewables. Mitchell (2008, 206) says that, although a given sustainable energy technology should be delivered by the cheapest means that is practical:

> [T]o reach a sustainable solution for the country it is necessary to break the band of iron, and that will require a shift of the political paradigm to one where the value of economic dominance is diluted so that, in matters of climate change, the environmental options takes precedent. That means as far as possible, when designing a policy, that the part of the economic dimension which slows the process down or limits innovation or change, should be by-passed (or replaced by 'just do it').

Kuhn's notion of paradigms is especially pertinent to the issue of how values interrelate with notions of rationality. Kuhn (1970) argued that different scientific paradigms were based on incomparable ('incommensurable') systems of language and values. Similarly, in the case of energy, judgements about whether we should aim for a 'renewable', 'nuclear' or fossil fuel future (or some combination thereof) are based on values. Arguments about the future (and often even current) costs and resources of different energy technologies will be heavily conditioned by the values underpinning the paradigms to which a given actor/interest group is affiliated. This is a fundamental point about renewable energy policy, since it is a matter of public debate and controversy and not merely a matter for discussion in boardrooms of multinational corporations where the main concern is (short-term) corporate profitability. Almost by definition, in the case of renewable energy, ecological rationality does (and should) come before economic rationality. Blühdorn (2000, 219) attacks EM for its (positivist) assumption that such judgements can be decided by science rather than through choice of values.

This criticism of mainstream EM's notion of ecological and economic rationalities being treated on an equal footing can be related to a criticism of mainstream EM's analysis of consumption. According to Spaargaren (2000, 58–9), consumption of services such as energy and water is 'routine'. Ecological change can occur after such services are 'de-routinized', for example, through high bills or concern about pollution. Spargaaren says: 'The ecological modernization of domestic consumption can be conceived as a series of de- and re-routinizations with regard to a great number of domestic routines or domestic practices, ranging from child rearing to gardening or doing the laundry. When elaborating upon the role of actors in the process of de- and re-routinization the concept of lifestyle of individuals seems relevant' (Spaargaren 2000, 59). Under the mainstream EM narrative, environmentalist pressures occur to correct the sources of the episodes of 'de-routinisation', and this process is completed as the relevant industry introduces practices that remove the sources of the problem. However, the likelihood that this leads to sustainable outcomes is challenged by Shove (2003). She doubts that consumers will re-evaluate what they regard as 'comfort, cleanliness and commitment'…'unless environmental considerations prove weighty enough to displace contemporary understandings of what everyday life should be like' (Shove 2003, 8). Shove points out that, while environmental considerations may 're-routinise' certain aspects, this is likely to be outweighed by continued 're-routinisations' to accept increased standards of 'comfort, cleanliness and convenience' according to changing conceptions of what is normal life. She comments that: 'programmes designed to improve the technical efficiency of air conditioning systems or freezers inevitably but perhaps inadvertently help maintain and sustain ways of life that include and increasingly depend upon artificial cooling. In this respect there is a close – but often problematic – connection between such initiatives and the dynamics of practice and escalating consumption' (Shove 2006, 32).

In the case of building regulation in the UK, for example, energy efficiency standards for new buildings have been progressively tightened. However, buildings with air conditioning will, in effect, be allowed to consume more energy than buildings without air conditioning. Air conditioning units will have to meet standards of efficiency themselves, but the acceptance of air conditioning as a 'normal' feature of buildings even in the relatively cool UK is perhaps a good example of the point that Shove makes about how trends towards new electricity uses being acceptable for buildings are undermining drives for more energy efficiency.

Is ecological modernisation therefore doomed to merely postponing the inevitable unsustainable outcomes of industrialised economies? Certainly, perhaps, 'yes', if ecological modernisation is seen as being pretty much solely a theory of production involving a usually passive consumer who leaves technology decisions to mainstream industry (in this case the energy and buildings industries). The argument in this book presents a tentative 'no' answer to the question, provided that the 'normalised' patterns of consumption are disrupted on a permanent basis by new, greener, technological identities. It is an argument for an alternative, albeit still technologically oriented, notion of ecological modernisation that hinges on mobilising support for green technological identities.

Identity ecological modernisation

Questions of identity have been very prominent in the minds of social movement theorists. Writers such as della Porta and Diani (1997), Melucci (1995; 1996) and Offe (1985) have charted how 'new' social movements have focused on identity issues, cultural struggles, non-material concerns, compared with the allegedly more material concerns of 'old' social movements. According to Light (2000, 60):

> Very generally, one can describe a politics of identity as a politics where agents ground their self-conception as political agents in some aspect of their identities. Often this identity is defined

negatively: one is marked by social forces as possessing an identity trait and then subject to different forms of oppression as a result of that trait.

In this case, if the 'agents' are ordinary energy consumers they may support a 'politics' of instituting incentives for renewable energy (expressed often as support for particular technologies). In the past, at least, this has often been grounded by a desire to see themselves as being 'anti-nuclear', and more recently as being 'against global warming' or 'self-sufficient'. However, Light, like others who discuss ecological identities, usually see this identity in 'non-material' terms (de Groot and Steg 2008). Yet positive popular identification with solar power or wind power or other renewable technologies is a hybrid of the material and the non-material. It may spring from non-material values in the mind of the energy consumer or environmental activist, but it ends in a material and technological solution through support for financial investment in renewable energy business activity. Through this means the renewable energy industry is associated with an ecological identity. Individuals can even pursue an ecological-energy identity by installing solar panels on their houses. As will be discussed in Chapters 3 and 4, in the 1970s supporters of renewable energy and 'soft energy paths' espoused non-material objectives in order to advance an alternative energy paradigm to that presented by the pro-nuclear establishment. Wellock (1998), cited in Chapter 4, talks about the tradition of concern for 'non-material' concerns in California that became associated with the anti-nuclear and alternative energy movement. It is common these days to see business trying to exploit associations with ecological identities with products that are 'green' or 'greener' in some way. However, renewable energy is an example of a whole industry, really groups of industries, that owes its (their) very existence to such an identity. It may not be the only technology that can claim this – organic agriculture, for example, associates itself with, and mostly builds itself from, an ecological technological identity. It may be that this 'identity ecological modernisation' approach may serve the organic example too, although further consideration of this question will have to wait for another project.

Hitherto, with notable exceptions such as Jamison et al. (1990), there has been little attempt to develop an analysis of how green social movements have approached technology issues in a positive sense. Social movements portraying a negative rather than positive ecological association with technology have been the focus of discussion that has been given most attention by sociologists of risk. However, there is an alternative to the bleak possibilities of either being in opposition to industrial technology, the so-called de-modernisers such as Sachs (1993) and Shiva (1989), and those 'mainstream' EM advocates such as Huber (1991), Mol (1995) and Spargaaren (2000), who seem to argue that the choices about what technologies are implemented to deal with ecological problems are pretty much solely made by conventional industries.

Jamison (2001) has raised the issue of how social movements have, at certain times, adopted identities associated with technological innovation. Jamison talks about how the environmental social movement has formed a 'collective identity' consisting of a 'cognitive praxis' made up of 'world view assumptions (cosmology), criteria for technical change (technology), and organisational forms (organisation)' (Jamison 2006, 47). We can simplify this further when we talk of a process called 'identity ecological modernisation'. This notion of identity can apply at the individual or much wider levels.

As is discussed in later chapters, notions of energy independence are often accepted as part of a region or country's cultural identity, and renewable energy is seen as a clean way of achieving this. Renewable energy emerged as an alternative to nuclear power in the 1970s and 1980s as part of a reactive identity of a movement seeking to assert itself as promoting an alternative. This identity has often been associated by pro-renewable campaigners with a state's identity of energy independence, as will be discussed in later chapters in the cases of Denmark and Texas. This concern for energy autonomy has been championed by Scheer (2007). In recent times in particular, individual energy consumers (especially in Germany) have pursued an ecological-energy identity through installing solar pv panels.

One avenue for an ecological modernisation strategy, which seeks the 'greening' of production and consumption rather than its abolition, is to insinuate itself into the normality of consumption by acquiring an identity that makes it part of consumption rituals. It is no contradiction to combine the concepts of 'normality' and 'rituals'. Douglas and Isherwood (1979) comment:

> Consumption uses goods to make firm and visible a particular set of judgments in the fluid processes of classifying persons and events. We have now defined it as a ritual activity...Within the available time and space the individual uses consumption to say something about himself, his family, his locality, whether in town or country,

on vacation or at home...Consumption is an active process in which all the social categories are being continually redefined. (Douglas and Isherwood 1979, 67–8)

A limitation of the 'mainstream EM' approach is that, using Mol's case of the chemical industry, it can leave the resolution of environmental problems to the production side, to be dealt with by the conventional industry through the sanitisation of its production processes and product performances into becoming otherwise normal products able to carry out their ritual tasks. The problem is that the products become re-routinised and simply continue to perform these rituals, for example, houses using paint made from less polluting materials and processes, air conditioning units designed to achieve greater efficiency, but without any sign that the consumer is consuming a 'greener' product. As part of counterbalancing the new energy-using consumption activities such as air conditioning, new 'ecological' rituals need to be inserted into consumption patterns; buildings in the EU now need energy performance certificates, although, as in the case of commercial buildings, perhaps these need to be clearly visible. What matters is not so much a transfer of energy information as the marking of houses as having more or less ecological status, thus conferring more or less status in this regard on the consumer. This type of incorporation of green rituals into consumption patterns has already been achieved to an extent in recycling activities, where now people are expected to play their part in recycling their rubbish according to the standardised norms – it has become 'ritualised', but it is an ecological ritual. People must now attain an ecological identity if they are to be regarded as performing the ritual patterns of consumption expected of citizens. In the energy field itself, the marking of refrigerators for energy efficiency according to an ABCD categorisation to mark their relative energy efficiency has been a success in improving the energy efficiency of fridges, the main problem being that the labelling system has not kept up with the improvement in energy efficiency. Consumers have taken considerable notice of the scheme, even though they do not understand the technical details; they just assume that it is the accepted thing to choose an 'A' grade fridge if possible (McDonald et al. 2007, 36–7).

We can still have our ecological modernisation, including economic development with ecological improvement, but only if the development is made conditional on giving primacy, not equality, to the ecological rationality. In this example this would mean that improvements in services must be accompanied by a reduction in carbon emissions.

Installation of renewable energy could be one of the pathways in achieving this, however expensive or allegedly economically unviable some may claim this to be. To use Mitchell's (2008) phrase it should be a question of 'just do it'.

The desire for solar pv technology by increasing numbers of households in some countries may reflect a desire to conform to an ecological identity. Being green is an increasingly important 'ritual' to which people aspire in the West. Sporting solar panels is an effective way of displaying that identity. Of course, if one simply combines an ecological with an economic rationality, one usually comes to the conclusion that various energy efficiency techniques are the first choice for adoption. However, such things do not afford the consumer any visible claim to an ecological identity, which goes some way towards explaining why energy efficiency seems less glamorous and often less popular than the 'expensive' solar pv panels. People can see solar panels, and can see your association with a green, self-sufficient identity. However, people cannot see loft insulation. The general positive popular identification with renewable energy coexists with more individualistic drives to identify with renewable energy. In some senses it might be possible to compare the desire to invest in solar panels with the type of status-seeking 'conspicuous consumption' that Veblen (1970) identified as being an explanation for the spending patterns of the well-off.

Energy efficiency advocates are sometimes upset by what they see as a diversion of investment on expensive solar pv when money is not being spent on improving the energy efficiency of the building. They are not alone. Executives of conventional energy suppliers are quick to point out that a 'rational' investment plan in low-carbon technologies would spend a given investment fund on a range of technologies cheaper than solar pv. However, an analysis based on an appreciation of an identity approach to ecological modernisation suggests that this transfer of investment would not happen. Energy consumers may tolerate (some) increases in their bills to support low-carbon energy technologies to the degree that they identify with them as desirable. The relatively low scores of, say, nuclear power compared with solar power in opinion surveys indicates that often the public would prefer to spend money on solar power, and also wind power and some other renewables, before nuclear power, and indeed carbon capture and storage in fossil fuel power stations (Borchers et al. 2007; Shackley et al. 2005). The point here is that it is not much good arguing that solar power subsidy programmes should be spent on something else if the solar subsidies are not transferable. The support for solar pv incentives is based on consumer identification with what as seen as a positive technology, not simply that it is a 'low-carbon' technology. Without this positive identity the subsidies, which come from energy consumers' pockets, would simply not be available for other purposes.

Hence the suggestion that energy efficiency be promoted by affixing clear notices to buildings (including domestic residences) that have acquired high efficiency standards so that the energy consumers can enjoy the heightened sense of ecological identity that the ritual of visible accreditation would deliver. All of this suggests more grass roots consumer involvement and mobilisation in technological implementation than the mainstream EM strategy would imply – that is, if it is interpreted as a strategy that depends on action being largely left to the conventional energy industries. In theoretical terms we need an EM strategy that involves less concern with balancing ecological with economic rationalities and more concern with promoting ecological identities and their associated ecological technologies.

Other, more 'radical' EM theorists, such as Hajer (1995) and Christoff (1996), suggest versions of EM that imply more bottom-up versions of EM. Christoff (1996) talks about giving less emphasis to material consumption priorities. Should I not use these 'radical' EM critiques to analyse the renewable energy case study rather than describing an 'identity' EM version? I argue that this is not plausible because the 'radical' EM theorists do not deal adequately (for the purposes of this study on renewable energy) with the inadequacies I have identified with the mainstream EM school, and moreover they do not provide a sufficiently positive (as well as oppositional) attitude to technology. I now justify these assertions.

The 'radical' EM school

Just as my argument agrees with some parts of the mainstream EM analysis, but disagrees with others, my argument complements some parts of what is here called the 'radical EM' school, but diverges on some key issues. In contrast to mainstream EM theorists, Hajer (1995) focuses on the role of social movements in environmental politics. Hajer's comparative account of the politics of acid rain abatement in the UK and The Netherlands has been widely applauded for its use of discourse analysis of how environmental problems are constructed and his study of the gaps between discourse and practice that occur. He criticises mainstream EM as being, in effect, to use Christoff's (1996) categorisation, a 'weak' form of EM. Hajer stresses the need for active engagements of social movements in decision-making and proposes the establishment of formal mechanisms for deliberation.

According to Carter's summary of Hajer (1995):

> in its weaker 'techno-corporatist' form, ecological modernization focuses on the development of technical solutions to environmental problems through the partnership of economic, political and scientific elites in corporatist policy-making structures... The stronger 'reflexive' version of ecological modernization adopts a much broader approach to the integration of environmental concerns across institutions and wider society which envisages extensive democratization and concern for the international dimensions of environmental issues. (Carter 2001, 214)

Hajer attacks the mainstream notion of EM, implying that it is 'techno-corporatist' in nature. According to Hajer (1995, 281), mainstream EM seeks to create technocratic structures 'where the best people can work in relative quiet' to generate a language of universal answers. Hajer counterposes this 'techno-corporatism' with his own preferred notion of 'reflexive ecological modernisation'. He argues that social movements have an essential role to play. The institutionalised deliberation mechanisms he proposes will promote 'the importance of the mobilization of independent opinions versus the respected power of authorities' (Hajer 1995, 281). In many ways Christoff (1996) deepens this line of attack by posing a distinction between 'strong' and 'weak' EM. Weak EM is identifiably techno-corporatist, focusing on technology and highly prescribed top-down practices, while strong EM involves grass roots-based action and diversity.

My own analysis moves in sympathy with Hajer's sentiments cited above, that is, against seeing environmental policy-making being made by elite groups in relative quiet. I would argue that, if this logic is applied to renewable energy, then this general argument (in this case) would be correct in that debates about renewable energy policy are very public in nature. Of course this may be an artefact of different case studies, of differences between renewables and the chemicals industry. Even so, this undermines claims that mainstream EM theorists may have concerning the 'generalisability' of their hypotheses. In the chemicals industry, technology selection is usually done by the chemical industries themselves. Regulations are proposed and accepted for achieving environmental targets, and the industry itself can decide how to address these targets and requirements.

As will be clearer in the succeeding chapters, environmental groups and social actors of various sorts do not always just demand 'emission reductions' of various types and leave it up to the electricity/energy industry to deliver them (although this was the case with acid rain abatement, studied by Hajer). Public pressure forces incentive structures on particular technologies, with public pressure being to apply incentives at different levels for different technologies in accordance with different deployment requirements. Environmental groups tend to be less enthusiastic about some options than others. For example, NGOs like Greenpeace have lukewarm attitudes to carbon capture and storage, and they are traditionally very hostile to nuclear power. Such NGOs are much more enthusiastic about wind power and solar power. In this they tend to follow the contours of public opinion in their notions of what technologies they are willing to support through paying higher electricity bills (Borchers et al. 2007; Shackley et al. 2005). Environmental groups are certainly unwilling to leave such matters to the electricity industry.

Is Hajer (and Christoff) 'weak' on technology?

However, the approach within this book also diverges from Hajer (1995) in several important respects. One of these concerns technology. This book's analysis is centrally concerned with the development of technology – renewable energy technologies – but Hajer (1995) and Christoff (1996) tend to see the technological focus of mainstream EM as a matter for criticism. They imply that there is a conflict between concern for technological issues and a bottom-up, participatory strategy. Indeed, an argument in this book is that EM strategies in the renewable sphere often depend heavily on positive grass roots and public participation engagement with renewable energy technologies. I would agree with Mol (1995, 43) when he says that 'the general emphasis on the importance of the influence of technology in socio-ecological transformations has remained a feature of ecological modernization theory.' I would subscribe to this on both analytical and normative levels, although I disagree with the emphasis that Mol places on achieving technological transformation by virtually complete reliance on the conventional industrial establishment.

Hajer, for his part, places little emphasis on discussion of technological issues. He notes a contrast between 'intermediate' technologies and decentralised communities in the 1972 publication 'Blueprint for Survival' and, on the other hand, 'technological fixes' in the 'Limits To Growth' report (Hajer 1995, 80–7). However, from there on any discussion of decentralised technologies appears lost in much wider discussion about political decentralisation. There is a lengthy debate about changing environment NGO tactics, different notions of democracy and political participation and the way that environmental problems are constructed. For example, Hajer says: 'reflexive modernization would seek a strengthening of public, inter-discursive forms of debate in order to make environmental politics a matter of deliberate and negotiated social choice for certain scenarios of societal modernization... [A]n extension of the possibilities for open deliberation would help find the socially acceptable strategies of modernization that at the same time might produce better results in terms of problem closure' (Hajer 1995, 282–3).

There are two issues here. First, however important deliberation may be, it does not in itself generate technological solutions. There is some discussion of this issue in Chapter 3, when I analyse arguments about planning consent for wind farms through a frame of 'conflicting identities'. Second, Hajer says very little about the involvement of environmental NGOs or environmental activists in promoting technological solutions themselves, or how initiatives initially dubbed by some as 'intermediate technology' have developed into mainstream renewable energy technologies. Indeed, by failing to emphasise the evolution of technologies (such as renewable energy) as distinctive new industries, often arising in conflict with the existing industrial establishment, Hajer wrongly gives up a lot of argumentative territory to that of the allegedly 'techno-corporatist' mainstream EM. Hajer mentions Foucault's scepticism about scientists, and declares: 'Reflexive modernization emphasises the importance of the mobilization of independent opinions versus the respected power of authorities' (Hajer 1995, 282), but what perhaps he should be saying is that 'we should be emphasizing the importance of the mobilization of independent technologies versus the respected power of conventional industries.'

Hajer is thus open to the charge that he focuses on discursive form rather than technological substance. He focuses most of all on political tactics and deliberative structures rather than on the actual objects of policy implementation – that is, what technologies will be deployed, how they are going to be deployed, and the policy debates that surround such issues.

Part of an alternative path to this charge, and also a way of overcoming the criticisms of mainstream EM, is for EM narratives to analyse the involvement of social movements and ordinary technology users in the development and proliferation of the technologies, and also to analyse how, later on, the renewables industry has developed often as an independent industry with its own identity compared with the conventional electricity industry. Even though renewable energy industries may be organised as conventional industries today, the green movement still intervenes in debates between the different energy industries, to support renewables against fossil fuels and nuclear energy. We have to remember that it was ecologically informed activists in Denmark and elsewhere (as mentioned in Chapter 3 and in greater detail in Toke, 2011) who virtually invented wind power and other renewable technologies in their modern form. In Chapters 4, 5, and 6 there is analysis of how the renewables lobby contests policy with the conventional energy industry, often in alliance with environmental NGOs. If all that the green movement did was to criticise technology and leave organisation and positive choices of technology up to the established patterns of conventional industry, then this might lead to very limited ecological developments of technology.

It is one thing to criticise a notion of EM that is focused too heavily on elite decisions about technology, but Hajer runs the risk of implying that it is better to build institutions for deliberation rather than focusing on technology. 'Technology' is not necessarily the opposite of 'social movement'. There are key periods when these are combined, or (as now) at least aligned, and it may be that there are often problems of social movements not being sufficiently involved in 'doing' positive things about technology, not (merely) that there is a lack of deliberation.

Christoff (1996, 496) also takes his cue from Beck (1992) and others in arguing for 'a critical or reflexive relationship to certain technologies'. As is the case with Hajer (1995), Christoff (1996) gives little attention to the involvement of environmental NGOs in the positive promotion of technological solutions. An impression is given that technology is effectively ghettoised as being associated with 'weak' EM. For example, he comments:

> Insofar as EM focuses on the state and industry in terms which are narrowly technocratic and instrumental rather than on social processes in ways which are broadly integrative, communicative and deliberative, it is less likely to lead to the sorts of embedded cultural transformation which could sustain substantial reductions in material

consumption levels, significant and rapid structural transformations in industrialized countries and major international redistributions of wealth and technological capacity. (Christoff 1996, 489–90)

Christoff tabulates a distinction between 'weak' and 'strong' versions of EM, with a 'strong' version being necessary to achieve ecological sustainability.

This type of 'weak' – 'strong' analysis of EM is widely used among radical academic environmentalists, and, for example, Dryzek et al. (2003) see the solution to the failings of the 'weak', state-centred, 'technocratic' EM as being through the adoption of a dual strategy of state action and 'continued confrontation in the public sphere' (Dryzek et al. 2003, 193). As has been implied already, this is useful as far as it goes, but, like Hajer (1995) and Christoff (1996), this strategy fails to sufficiently emphasise the importance of a positive engagement of civil society in making technological choices – that is, as opposed to a trajectory solely of opposition to some technologies. Christoff appears to set up a division between 'technology' and 'discursive and participatory' modes, as seen in Table 2.1 and his discussion (Christoff 1996, 490), whereas an approach is needed that draws such strands together. A problem with Christoff, and also Hajer, is that, though their notions of ecological modernisation may be strong on culture, they may be weak in promotion of eco-technologies.

As we shall see, perhaps the 'heroic age' of grass roots involvement in actually inventing renewable energy technologies such as wind power and biogas has passed, but, as will also be discussed, the renewables industry is still battling for incentives, often in the face of the interests of conventional fossil fuels and nuclear generation interests. This implies that the positive engagement of civil society to make technology choices

Table 2.1 'Types of ecological modernisation'

Weak EM	Strong EM
Economistic	Ecological
Technological (narrow)	Institutional/systemic (broad)
Instrumental	Communicative
Technocratic/neo-corporatist/closed	Deliberative/democratic/open
National	International
Unitary (hegemonic)	Diversifying

Source: Table 1 (Christoff 1996, 490).

is still important for a strategy based on ecological modernisation. On the one hand, in the case of solar photovoltaics, direct social movement involvement including user-generation is still very important, and, on the other hand, avowedly big industrialist renewable trade associations lobby governments for incentives for large-scale wind power with the support of NGOs against the indifference or resistance of energy utility incumbents.

There is a danger that the green critique of technology can sometimes be interpreted as taking a sceptical stance towards all technologies. Fudge and Rowe (2001, 1453) citing Beck and Lash, argue that 'A dominant (technologically driven) form of rationality may stifle the creative ways of thinking and living which the future will demand.' Beck (1992) analysed the nature of the technological scepticisms that formed the background to the rise of the green movement in the early 1980s in Germany. However, he said very little about green activist pressures for positive alternatives to technologies such as nuclear power that were the focus of the attention of the new politics of risk. Writers such as Touraine have also tended to focus, when discussing social movements, on opposition to technology rather than support for positive alternatives (Touraine et al. 1983).

Jamison (2006) maintains that social movement theorists tend to focus on issues of identity or access to political resources rather than relations with technology. Meanwhile the science of technology studies (STS) literatures tend to focus on the technology and treat the social movements as a 'marginal....amorphous background' (Jamison 2006, 45).

There is no fundamental problem with using 'identity' politics to analyse renewable energy and business strategy as part of ecological modernisation. The problem with the 'identity' concerns that have so far been the principal analytical focus is that the identities have tended to be seen as being in opposition to technology. Some innovation theorists have focused some attention on the role of outsiders in technical developments (van der Poel 2000), and Smith (2004; 2007) has focused attention on the importance of niches in promoting ecological technological developments. However, these types of studies, as Jamison observes, have not dealt much with the politics of the social movements, or identity issues. My concern here is to bring conceptions of identity and technology together, rather than seeing them as separate, even opposing tendencies. This work tries to focus more attention on how the renewable energy industry has constituted a separate industrial identity behind which a coalition of interests, including a social movement, has assembled to support its progress.

We need to outline a method of analysis in order to mobilise the 'identity' EM framework. As was discussed earlier, Mol uses a focus on interest group analysis of bargaining both within industry and also between environmental interests and industry. Hajer (1995), by contrast, privileges discussion of discourse over interests. Perhaps we need a balance between these approaches, one that merges discussion of both industrial politics and cultural politics. This can be discussed in the context of a critique of Hajer's use of the discourse analytical tool.

Discourse and interests

Hajer (1995), in combination with other social scientists such as Fischer (2003), has been highly influential in advancing discourse analysis as a tool and also as an epistemological approach in public policy analysis. Discourses, argues Hajer (following a Foucauldian line of argument), need to be the central point of analysis of how interests are constructed, how environmental policy changes are negotiated and how policy outcomes transpire.

Hajer employs the notion of 'storylines' to analyse how interests are constructed. He argues that, rather than view how interest groups enter into coalitions with those who have shared interests, we should look at how particular storylines effectively constitute the interests of those he describes as being in a given 'discourse coalition'. Hajer positions himself 'against studies that see social constructs as a function of the interests of a group of actors' (Hajer 1995, 59) and argues for his notion of 'discourse coalitions' which involve storylines, actors and practices and 'whereby storylines potentially change the previous understanding of what the actors' interests are' (Hajer 1995, 65–6). Analysts such as Szarka (2007) have used the 'storyline' metaphor to describe pro and anti-wind coalitions.

This approach has considerable advantages, but at the same time it is doubtful that it is possible to abandon all use of the notion of independent interest. Studies that analyse interests only as a function of discourses may only rely totally on discourse analysis if we can study all the multitudes of discourses involved in constructing an interest. The usual response to such a criticism may be that discourse is the only firm evidence we have about what constitutes interest. I do not wish to contest this proposition. Rather, we need to understand that in a given social situation there are many discourses acting to construct actors' beliefs about their interests, and that, if effective analysis is to be done, it is reasonable to make assumptions about what an interest could be for the purposes of the analysis. Otherwise the type of analysis that is possible is limited, not least because of a lack of resources in accessing and interpreting all of the (often conflicting) discourses.

For example, if we compare the discourses of renewable energy developers in the USA and continental European countries such as Germany on the topic of incentives for renewable energy (Chapters 5 and 6), there is a striking difference in the dominant discourses used by these renewable energy developers. Developers in the USA talk positively about trade in 'renewable energy credits' to support RPS policy instruments, while German renewable developers attack proposals for equivalent 'free market' mechanisms. In the USA, where there is a dominant 'free market' discourse, renewable energy interest groups may find their proposals more acceptable if they are presented as 'free market' instruments, while in Germany, where dominant discourses are different, this is not necessarily the case. 'Feed-in tariffs', involving renewable energy prices set by the government, are their preferred mechanism.

Are we simply content to assume that that the European and US developers have strikingly different notions of self-interest as regards financial support mechanisms? Or are we able to look beyond discourses and investigate how the details of renewable energy financial policy work out in practice? Could it be that the interests of EU-based and US-based renewable developers are closer than we might assume from restricting ourselves to a study of discourses?

In order to escape from this problem of assuming that discourse must supersede the notion of interest, we need to have a conception of interest that transcends public policy discourse. Then we can begin to unpick what is most necessary for effective renewable energy policy mechanisms. In fact, in the case of companies investing in renewable energy plant, renewable energy developers on both sides of the Atlantic will be subject to much the same demands from financiers for reasonable assurances of long-term income streams for energy generated. In this sense the developers in the different cases have the same interest, but use different policy discourses to obtain incentive structures that satisfy their common interests for long-term security of income flows.

We can analyse this type of situation by using the concept of interest as a 'short cut'. However, such a 'short cut' may nevertheless be a more thorough form of analysis than selecting particular discourses in isolation from many others that are relevant. As has been stated, in many situations there will be insufficient resources available to analyse all discourses involving a specific actor. In such situations it may often be necessary to assume an 'interest' that is distinct from the dominant policy storyline identified for a policy area in order to better understand policy outcomes.

It is necessary to be pragmatic about the use of discourse and interest group analysis. Sometimes it is useful to see interest as constituted by discourse, but sometimes it can be useful to analyse interest as being independent of discourse. Discourse analysis remains important, especially in cases such as pinpointing trends and changes in government policies, in discussing how identities are constructed and also in analysing how different political economies can shape incentive structures. Different dominant discourses (about political economy) can advantage or disadvantage renewable interest groups who try to interpret them to suit their interests. As Toke and Lauber (2007) argue, for example, 'neoliberalism' has been interpreted differently in Germany and the USA/UK. In Germany feed-in tariffs were promoted as a measure to help small companies against the interests of big monopoly energy utilities. However, neoliberalism is interpreted differently in the USA, and the idea of government setting prices (for renewable energy in this case) has been associated negatively with so-called 'command and control' policies. Is this dominant discourse a factor that limits the development of renewable energy in the USA? Renewable energy policy in the USA is discussed in Chapters 4 and 5.

Institutions

Analysis of discourse and interests needs to be conducted in the context of social structures. We can analyse relevant social structures through an institutional approach. An institution may arise from the pressure of traditions and the interaction with dilemmas, but once established the effects of the institutions (in terms of how actors react to incentives, norms and so on; see Scott 1995) will arise independently of the traditions that led to the establishment of the institutions. Hence institutions need to be studied separately from traditions if we are to understand how outcomes occur.

The study of traditions is important; for example, the tradition of environmental activism in California was important as a precursor to the anti-nuclear and pro-renewables movement, as is discussed in Chapter 4. As Bevir and Rhodes (2003) argue, traditions need to be studied in the context of how they evolve as dilemmas force actors to choose new paths, such as how anti-nuclear activists were induced to give (renewable energy) supply technologies more priority in the context of the 1970s oil crises, as argued in Chapter 3. Another example of how path dependence and tradition affected later outcomes is the tradition of rural self-reliance in Denmark, which led the anti-nuclear movement there to develop renewable energy technologies, especially modern wind power. However, traditions can be studied within the context of what Hall and Taylor (1996) call 'historical institutionalism'.

According to Hall and Taylor, historical institutionalists define institutions as 'the formal or informal procedures, routines, norms and conventions embedded in the organisational structure of the polity or political economy' (Hall and Taylor 1996, 938). We can trace political outcomes partly or wholly as a 'path-dependent' process. 'Institutions are seen as relatively persistent features of the historical landscape and one of the central factors pushing historical development along a set of "paths"' (Hall and Taylor 1996, 941). Hall and Taylor (1996) identify three types of institutionalism: rational choice institutionalism, which analyses outcomes by modelling how interest groups pursue their self-interests, often to generate outcomes that are suboptimal for society as a whole; sociological institutionalism, whereby belief systems underpinning activity in organisations are studied, which will tend to involve a broader definition of institutions to include 'frames of meaning' (Hall and Taylor 1996, 947); and historical institutionalism, which is said to involve a mixture of techniques, including analysis of rational choices, belief systems and, as mentioned already, 'path dependence' through historical analysis.

Institutionalism also exists in the form of 'economic institutionalism', as articulated by Hodgson (1998). He argues that markets are not best understood in some imaginary form whereby individual consumers make purchasing decisions purely on the basis of their own unique preferences. As is discussed further in Chapter 5 in connection with the example of incentives for solar pv in California, market outcomes are influenced by actors' pre-existing habits (whether followed consciously or taken for granted) and the way markets are designed. Consumer preferences are shaped by institutions, often based on widely held notions of status (diamonds, Rolex watches, French wine). Competition on price occurs in between the innumerable numbers of fissures represented by institutions. The outcome is only 'efficient' as far as the institutional structure allows. In the renewable energy case, outcomes are influenced by a range of state and Federal laws and policy initiatives, sometimes unrelated to renewable energy, but influential on producing a context that constrains the choice set of actions available to actors hoping to pursue their interests. This approach will aid understanding of the development of policies to support renewable energy and of how the interests of renewable energy groups interact with various institutions to generate policy outcomes.

Much discussion about the role of markets is impoverished by an assumption that market outcomes occur by some unknowable process (beyond classical supply and demand curve analysis), when in fact key influences on their outcomes can be analysed by study of the institutions that influence what can be produced and what can be consumed and what people want to consume. As is discussed in Chapter 3 in the topic of the social choice of technologies, what appears to be a 'successful' technology is only successful because, for example, its techniques and operation fall within the contours of existing social institutions and produce the outcomes that social groups want. Coal-burning technology was the prime choice for energy production for many decades, but was only successful as an energy technology until social groups' demands for cleaner air became more and more strident.

The concept of institutions may also be extended to cover perceptions of material artefacts. Although we cannot include material artefacts directly in social analysis, we can base analysis on actors' beliefs about structures in the material world. We may refer to them as 'material institutions', in order to distinguish them from conventional historical/ political, economic or sociological institutions.[1]

There is an important difference between seeing material institutions as just reflections of materiality and seeing them as social constructs. For example, in Chapter 4 there is a discussion of the 'wind rush' in California. This was influenced by the fact that the California Energy Commission had mapped wind speeds in parts of the state, an activity that had been done much more comprehensively than in any other state. Hence, even though California has much fewer wind resources than many other states, as far as the wind power developers were concerned the wind resources, for the purposes of commercial exploitation, effectively existed more in California than other places.

Policy networks can be described as institutions, and policy network analysis is certainly well used by Mol (1995). Indeed, there is coverage of policy networks and renewable energy in the case of the UK (Toke 2010). There is some reference to policy networks in Chapter 5 as a metaphor in understanding state-based decisions on renewable energy policy. However, I have avoided using policy network analysis as a general point of analysis because of the difficulty of comparing policy networks on the basis of the different national states relative to that of US states.

Conclusion

I have reflected on the opportunities and limitations of two forms of EM, which I have called 'mainstream' and 'radical', and talked about a third 'identity' approach, which is more relevant to the analysis of renewable energy development. I agree with the notion, implied by the mainstream version, that renewable energy needs to be implemented on an industrialised basis, but disagree in the case of renewable energy with its almost sole focus on conventional industrial regimes as the agent that is mainly responsible for making technological choices. The renewable case is better understood through study of how an emerging industry with its own political (and often engineering) identity, which is distinct from nuclear and fossil fuel technologies, has lobbied and fought to achieve an incentive structure that favours the renewable industries rather than conventional industry. In doing so it has formed distinctive coalitions with environmental and other citizen groups. Its early emergence is especially marked by an identity of being different from, often opposed to, conventional industry. Public support is mobilised behind this identity, and it is done on the basis of preference for technological solutions rather than merely estimations of cost-effectiveness.

In this sense 'identity' ecological modernisation can be said to emphasise 'sustainability' as a guiding force for 'development' in the concept of sustainable development. Mainstream EM often seems to do things the other way around, and to use development to guide a rather more limited conception of sustainability. Hajer (1995) and Christoff (1996) criticise mainstream EM for elitism and argue for more culturally oriented notions of EM involving popular debate and bottom-up political initiatives. However, they appear to sideline technological issues in favour of strategies focusing on different types of social and political organisation. While an increased focus on democracy and deliberation often has some merit, and bottom-up organisation can often be important (especially in the earlier innovatory days of renewable energy), a key point of the identity EM approach is that it is the public's choice of, and often involvement in, technological solutions that is at the heart of ecological sustainability.

This 'identity EM' narrative will be expanded and applied in the following chapters. I use the term 'narrative' here with purpose, in order to distinguish this EM account from the objectivist aspirations of mainstream EM accounts. Although narratives can use models to describe processes and relationships between actors, these models cannot be automatically transferred to other cases. Hence I would not claim, for example, that the 'identity EM' approach used here can apply to cases other than renewable energy. Indeed, not only is it the case that renewable energy may change so that the approach used here becomes less relevant, but also I develop a set of criteria to measure how the relationships between the actors may differ in different country-case studies at different times in the renewable energy case.

Only other empirical studies can throw light on how typical or atypical the two different case studies are (chemicals and renewables). However, even if the model suitable for renewable EM analysis were rare, it would represent a crucial example. Energy stands as a very important, some would argue as the most important, environmental issue, given the policy primacy given to the resource depletion and pollution issues surrounding energy use.

The analysis in this book draws heavily on the criticisms made by the 'radical' version of EM in the sense that discursive analysis of the public debates surrounding renewable energy is essential to understanding policy outcomes. This is especially relevant in this area, where pro-renewable outcomes have often been achieved through public debates and campaigning rather than quiet negotiations with the conventional electricity industry. Indeed, the renewable industry, as represented by new technologies such as wind power and solar power, would hardly exist today if not for the grass roots, often technologically engaged, efforts of public campaigners and technical artisans working outside the electricity regimes.

However, the analysis conducted in this book also departs quite considerably from the radical version on several counts. Most serious is the failure by Hajer (1995) in particular to engage more thoroughly with technological issues. In this case, this includes the ways in which NGOs and activists have joined forces with the renewable industry to develop and support specific renewable technologies. Risk theory is problematic in as much as it highlights opposition to energy technology, but says little about how social movements have aided the creation of new 'ecological' industries such as wind power, biogas and solar power. In addition, while Hajer's focus on discourse is important, the automatic privilege he accords to analysis of discourse rather than interests may obscure key points of understanding and render many important areas opaque to analysis. A method that involves both analysis of discourses and interests in the context of institutional analysis is utilised in this work.

As is implicit in the foregoing discussion, EM is perhaps not best seen as an attempt to divine an 'objective' account of ecological policy in Western states *per se*, but rather as a framework that provides sets of tools, categories and frameworks to generate narratives that are specific to particular technologies or places. If the aim of deploying 'identity' EM analysis in the various country-case studies is to be achieved, there is need of a tool that can help measure the degree of, and type of, commitment to 'identity EM' in each country-case study. Of course, the results of the measure may change in a given case over time. Hence a framework is set out here consisting of a set of characteristics of 'identity' EM that will assist in the measurement of EM in different instances.

Characteristics of 'identity' ecological modernisation

First is the significant contribution of idealism in support of particular renewable energy technologies. This is epitomised by support for renewables by activists operating from 'non-material' motives, for example to promote an alternative to the use of nuclear power or fossil fuels in order to avoid environmental problems associated with these options.

Second is the existence of dedicated financial support mechanisms for renewable energy technologies, which mean they can be deployed in large quantities. Feed-in tariffs are, arguably, the most successful type of support mechanism, although other mechanisms such as the British Renewable Obligation or the RPS systems in the US allied to the Federal production tax credit also qualify as financial support mechanisms.

Third, there are independent, publicly active, trade associations representing the main renewable energy technologies. Such associations may represent particular renewable energy technologies, or may act as umbrella groups for more than one renewable energy technology.

Fourth, the trade associations form coalitions with environmental groups to campaign for improved and maintained financial support systems for renewable energy technologies.

Fifth, much of the deployment of renewable energy is done by companies that are independent of the major electricity companies.

This list of characteristics is intended to act as a heuristic device. The point of a 'heuristic device' is to act as an approximation. We cannot compare different cases without using a common method of measuring them. We need to compare different case studies using a common set of categories, so producing a greater understanding of differences and similarities between case studies and also leading to a greater understanding about how outcomes occur. It can also pave the way for some generalisations to be made that can themselves be the subject of further discussion in other studies of empirical cases (Toke 2010 forthcoming). This set of characteristics will be deployed in the different country-case studies not only as a means of comparing the different countries, but also in measuring the ways in which EM has altered, or become more or less visible in different forms, at different times in a given country.

3
Renewable Energy: A New Identity and a New Industry

The aim of this chapter is to begin to explain the politics of the emergence of renewable energy as a major energy source since the 1970s. It follows on from the previous chapter in that a central aim of this chapter is to discuss how the identity of renewable energy sources has emerged, changed and developed in response to the social, political and economic context. The theme of identity is important to understanding a technological development that is distinctive from many or most other technological transitions.

It is a transition that could hardly have occurred without major state-sponsored restructuring of the incentives and regulations governing energy in favour of renewable energy. As is discussed elsewhere, this restructuring was begun and also, in various instances, helped along by a revolution from below, social movements and cause interest groups who supported the granting of special incentive structures for the nascent, and also existing, renewable energy industries (Toke 2011). Dryzek et al. (2003) highlight the importance of environmental movements to an EM strategy of setting environmental objectives, but the point being made here is, in the case of renewable energy, that popular pressure is important in selecting the technological means for achieving these environmental objectives.

The transition towards the mainstream adoption of renewable energy has been continued because the renewable industries have continued to find political sustenance for the incentive structures that support the industry from the positive identity of the renewable technologies. Certainly, there is a pattern of highly contentious planning disputes surrounding wind power in particular in many cases (in some states more than others). This aspect will be discussed later in the chapter. However, without the positive identity that wind power and other renewable energy industries enjoy, it would be impossible for the coalition of industrial and environmental groups that support renewable energy to be successful in securing and maintaining the incentives and regulations that allow significant quantities of renewable development.

Following a brief discussion of the early history of electricity and the beginnings of wind power as a technology, there will be a discussion of the discourses that sparked development of renewable energy sources, and a study of how renewable energy was, in the 1970s, associated with anti-nuclear movements. The identity of the 'alternative' anti-nuclear energy movement will be discussed and there will be an analysis of how the notion of 'decentralisation', frequently adduced to support alternative energy strategies, fits in with conceptions of renewable energy identity. There will be a discussion of the emergence of some key renewable energy technologies in a modern form, paying most attention to wind power. As mentioned earlier, there will be a discussion of how positive renewable energy identities can conflict with other identities. Finally, there will be an analysis of how the earlier visions of energy paths compare with contemporary discourses on the role of renewable energy.

Renewables – a solution waiting for a problem

Wind power, which now provides the bulk of new renewable energy supplies around the world, has existed as technology able to supply electricity for more or less as long as electricity has been supplied commercially. Edison first supplied electricity for lighting to a New York apartment block in 1882. This was supplied by a coal-fired 500 kW steam engine linked to a dynamo.

We should not forget that hydroelectricity has been a major renewable energy supplier since the beginning of the commercial electricity system. However, variable or intermittent renewable energy supplies do not feature much in the history of the development of the electricity system (until recently of course) – indeed, Hughes's (1983) classic account of the development of electricity systems in the West does not even mention wind power once. Yet electricity was generated from wind power from the early days of commercial electricity. James Blyth, a Scottish academic, was arguably the first to generate electricity from a wind turbine in 1887 (Price 2005) and others followed suit soon afterwards. There was even a cottage industry making electricity-generating wind turbines for use in rural off-grid situations in countries such as Denmark and also in the US Midwest (Hau 2006; Righter 1996).

The point that I make here is that wind power existed as a techno-logical option. However, to use social of construction theory (SCOT) (Bijker 1995), it was not that wind power was a failed technology, but rather that the choices of social groups did not favour its widespread adoption. Bijker argues that the success of a technology is dependent on the extent to which there are social groups who feel that it satisfies their aspirations. Technologies are chosen for their perceived social uses, not for their inherent technical qualities. Until well after World War Two, there was little questioning of the dominance of the fossil fuel industries, whose use expanded massively during this period. However, their success was predicated largely on the fact that the dominant social groups and social demands of the day found that the technologies fitted in with what they wanted, not because the fossil fuel technologies are inherently successful on technical grounds.

Hence, apparently obvious stories of technologies that appear successful because of their inherent technical characteristics look incomplete when viewed through the lens of SCOT. For instance, it may be argued that coal became the prime choice for energy production for many years because it was low-cost and very abundant. Yet we should really mention that the growth of industrial production, urban living and demands for improved living standards formed the social pressure for coal use. We should also mention that for many decades the sort of institutional constraints on coal use that exist at present were not in evidence during the days when coal was 'King'. There were few social pressures (that were politically effective) for constraints on coal use for environmental purposes. It was used in all sectors, domestic, industrial, and for electricity production. However, social groups emerged that campaigned for clean air, later on against acid emissions, and now on climate change. The pollution problems associated with coal led to regulations that were restrictive on coal use, or that increased the costs of coal burning relative to other options. The social pressure against pollution from coal also tipped the scales in favour of public regulation and often investment in other energy technologies, particularly natural gas, which is seen as a much cleaner option.

Nuclear power has been unable to do as much as was once thought to take advantage of this position (constraints on coal) because of what became, from the 1960s to the 1980s, increasingly stringent requirements for plant safety, which made the plant more expensive to build. The impact on coal, for example, included more and more requirements to restrict emissions of particulates, sulphur, nitrogen oxides, heavy metals and waste.

Nuclear power also saw its costs increased, not so much through having to file planning documents as in the changes that were needed to power plant design itself. These increased construction costs. Some of the factors that made nuclear power more expensive were the outcome of new regulations applied to all power stations, for example, regulations to protect fish from the effects of power stations drawing in water. For example, in the US, because of 'the risk that they will be ordered to perform with the cooling water intake structures at...zero kills [of fish], prudent construction [by developers of power stations] has by and large avoided constructing anything [power stations] that draws in new water' (unattributable interview with leading renewable energy executive and activist in Texas, 1 May 2009).

Use of waters was one of many issues for which environmental lobbyists sought, though rule changes at various levels, to internalise external environmental costs. In terms of the many law cases brought by environmental lobbyists in the 1960s to the 1980s, 'most of the litigation was about trying to get the government to do its job, which had the economic impact of internalising these externalities, or at least coming up with the pricing regime in effect, a costing regime in effect that the regulation creates. The big one on the table now is carbon' (unattributable interview with leading renewable energy executive and activist in Texas).

What I do here is reinforce the SCOT message that the issue of whether nuclear power or coal power or gas power (or wind power) or whatever is or is not a successful technology depends on the (changing) pressures of social groups. Indeed, the study of institutions can be seen as a link between the demands of social groups and outcomes. As the Texan lobbyist quoted above implies, changes in market costs and prices are generated through changes in rules and regulations. To speak of 'market prices' ignores the fact that markets are themselves constituted by innumerable institutions, many ensconced in regulations, some by convention, and these social pressures aggregate to define the extent to which a technology is 'successful'. Certainly, the nature of the technical artefact is of central importance to what is possible in the energy world. Indeed, much is written in technical papers and books about such things. Much less is written about the other side of the technology coin, that of the politics of technology, of which Bijker (1995) has spoken. One of the aims of this book is to right this imbalance in the case of renewable energy, and specifically link the 'political' analysis to political science theory itself.

In the case of wind power, to use the SCOT formulation, before the oil crises of the 1970s the only social groups that might have supported the (wind power) technology were those who wanted electricity for off-grid purposes. Wind power was simply not seen as a serious method of supplying power to the grid, the development of which increasingly became seen as the electrical imperative of the day for governments and system planners. Coal supplied most electricity in the industrialised world, although in some countries, including the USA, hydroelectricity was another major supply source. It was only after World War Two that oil and gas began to supply a large proportion of electricity, although even in the USA coal has continued to supply around half of total electricity. An important social group that wanted wind power was lacking; that is, until the emergence of the anti-nuclear movement and its need for an alternative to nuclear power in the context of the oil crises of the 1970s.

It is relevant here to introduce a concept drawn from institutional theorists such as March and Olsen (1989), namely the 'garbage can' approach. Essentially this means that problems and solutions are independent of each other and connect together usually by a random process of coincidence.

Wind power, as an option for supplying electricity networks, used to be a solution that did not have a problem with which it could connect. As long as there was thought to be sufficient coal, or other fuel supplies, wind power, which remained a very underdeveloped technology, was not seen as a sufficiently serious option to warrant attention by governments or electricity utilities. Western government programmes were concerned with developing more comprehensive grid systems. Wind power was saddled with an identity that seemed old-fashioned, unsuited to the demands of modern industrialisation for power that could be efficiently and cheaply provided.

A few enthusiasts developed prototype machines, but the very arguments deployed in the technology's favour seemed to chase an identity to which it could not easily aspire. Palmer Putnam, who built a 1.25 MW two-bladed machine (which worked only sporadically) in Pennsylvania in 1939, commented in his book that his wind power project 'is directed towards anyone interested in man's instinctive urge to subdue and harness his environment' (Putnam 1948, vii). This statement, today, appears almost as a stereotype of what is bad about unsustainable industrial (macho) attitudes to nature! However, in its context it can be seen as part of an unsuccessful attempt to develop an identity that fitted in with dominant discourses on the role of energy sources and energy supply systems. Heymann describes declining German utility interest in R&D in wind power at the end of the 1950s, and says: 'Post-war energy problems had been solved and fuel prices were rapidly falling. Additionally, in Germany and the USA, nuclear power was embraced as the solution of future energy problems. Small and unstable generating wind turbines finally appeared ridiculous' (Heymann 1998, 119–20).

Yet it was the dominant attitude to nature that was to come under increasing attack, sporadically in the 1950s, but with greater force in the 1960s and 1970s. This change correlated with incipient concern with the issue of depletion of fossil fuel resources. Indeed, nuclear power, in 1950s a new fuel technology, promoted itself largely on the basis of such concerns, promising a new future based on rapidly increasing demands for electricity.

The story of the rise of the anti-nuclear movement is well known and well covered in the literature. Much has been written about the negative aspects of a technology increasingly beset by accusations that it was a secretive harbinger of a dangerous, radioactivity-polluted, increasingly authoritarian society that was subject to a growing social movement in opposition to its spread (Jungk 1979; Patterson 1976; Rudig 1990; Weart 1988). However, the new energy force had, for a long time, sufficient positive identity to attract support for special investment programmes in most industrialised countries. By the end of the 1950s there had been a growth in many Western countries of nuclear energy institutions, including Euratom, which was founded as a result of the 1957 treaty that accompanied the establishment of what became the European Union itself. Statements from governments, such as the UK Government in the mid-1950s, that urged investment in nuclear power justified this by deploying such sentiments as: 'Improved living standards...can only come about through the increased use of power. The rate of increase required is so great that it will tax the existing resources of energy to the utmost' (Ministry of Power 1955, 11). Although it appears from this quote that the overwhelming social demand was for more and more energy, this demand was, within less than 20 years, to come into conflict with other social demands to, as was discussed earlier, internalise environmental externalities of energy production.

It should be recalled that the solution and the interests concerned with 'nuclear power' remain today, although some of the discourse has changed. We can see here how interests can stay much the same but discourses can change, since today the discourse in support of nuclear power is likely to be based heavily on the reduction in carbon dioxide emissions (as well as the older aspiration to avoid the problems of fossil fuel depletion). This is one among various examples given in this book concerning how we need to study the interaction of interests with institutions to produce outcomes, rather than just relying solely on discourse analysis. In this case nuclear interests have shifted their discourse in line with changing social group pressures, away from a discourse of needing to meet ever-increasing demands for energy, and into a more contemporary social discourse of dealing with climate change.

Nuclear power was able to develop programmes in the 1950s and 1960s in countries such as the UK and the USA partly because of the structure of the electricity industry. The industry in both countries consisted of retail monopolies, which meant that, provided the nuclear power programme remained relatively unquestioned as a necessary bridge to the future, it did not matter very much how expensive the nuclear power stations were, since the consumer had no choice but to buy electricity from one supplier. The suppliers could add on the costs to the bill which the consumer would pay. However, the discursive consensus on the need for nuclear power began to crack, especially in the 1970s.

There is a paradox in the fact that nuclear power was fiercely opposed in the 1970s, given that the onset of the oil crises in October 1973 and in 1979 seemed to make the need for alternative energy supplies more crucial. In fact, since demand for electricity fell in this period and increased thereafter only at a reduced pace (compared with the 1960s), the electricity utilities became less keen on expensive new nuclear power plant, made more expensive by stiffened regulatory requirements (Komanoff 1981). This said, Wellock (1998) argues that in the case of California it was not economics but social movement opposition that ended the nuclear construction programme in that state. Here we see evidence for the importance of discourse analysis to study how outcomes occur. This is because the opposition to nuclear power needed a discourse of an alternative. It needed to project an energy identity that filled this gap. Initially, there was no significant renewable energy interest, but this emerged as something that benefitted from a positive discursive context, that is, the need for anti-nuclear activists to find a positive identity that could substitute for nuclear power. On the other hand, as we shall see in Chapter 4 on California in the 1970s and 1980s, we also need to analyse how interests interact and alter institutions if we are to obtain a better understanding of how outcomes occur. For now let us see how a new energy identity, a renewable energy identity, began to emerge.

Defining the energy alternative

Initially, in the 1960s and early in the 1970s, work published by the (often perceived as) conservative environmental group Sierra Club issued a 'plague on all your energy sources houses' approach, which put most emphasis on energy conservation rather than renewable energy as an alternative (Holdren and Herrera 1971). Criticism of, and calls to oppose, nuclear power stations were a key feature of the document, which attacked the vision of rapidly increasing needs for energy supplies that underpinned pro-nuclear power discourses. Solar power was seen as an important long-term resource, but renewable energy was not seen as a clear alternative strategy to nuclear power (Holdren and Herrera 1971, 142–3). Global warming from carbon dioxide emissions was important, but only as part of the notion of generalised thermal pollution from energy sources, which was seen as a major long-term threat to the global climate and ecosystems (Holdren and Herrera 1971, 88–98). Nuclear power was identified as a cause of this thermal pollution, which was widely cited by anti-nuclear writers as a threat. Ground-based (as opposed to satellite-based) solar energy was seen as practically the only possible energy source that did not cause this problem. Tidal energy was said to offer only a small prospect of energy supplies, and ...

> Wind energy is even less promising as there are few places in the world where the wind is strong enough and steady enough to make harnessing it for the large-scale production of power at all interesting. The ultimate source of the energy in the wind is the sun; the prospects for exploiting that directly seem much brighter. (Holdren and Herrera 1971, 118)

This approach placed relatively limited emphasis on renewable energy and laid considerable stress on energy conservation. This was mirrored by the 'Blueprint for a Green Economy' (Goldsmith et al. 1972) issued at the beginning of 1972, in combination with the first newsletter published by UK Friends of the Earth. There are many mentions of the need for measures that are 'energy conservative' and the need to reduce energy demand, but there seems to be no clear advocacy of renewable energy. In the case of Schumacher's 'Small Is Beautiful' text (Schumacher 1973) written just before the October 1973 oil crisis, you will search in vain for mentions of renewable energy. There is biting criticism of nuclear energy. 'It is hard to imagine a greater biological threat, not to mention the political danger that someone might use a tiny bit of this terrible substance (uranium) for purposes not altogether peaceful' (Schumacher 1973, 25). Conservation of resources and a non-material, spiritual approach to life are recommended, based on Buddhist and Gandhi-ist concepts. Intermediate technology, that is, modern technology suited to the needs and also pockets of local people, is recommended. Schumacher's name became associated with decentralised renewable energy, although initially his focus was on intermediate technology that could be used in the developing world (Schumacher 1972). Although the notion of decentralisation is frequently cited as being an important part of the 'identity' of the energy alternative movement to nuclear power, it is, as will be discussed, a rather amorphous and vague claim to an identity.

John Holdren (and the other authors discussed here) was writing with the benefit of what was then seen as a good grasp on energy environmental issues. Today he is seen as a world leading expert on climate change and is Barack Obama's Chief Scientific Advisor. It is also worth pointing out that, although global warming was mentioned by Holdren and Herrera (1971) (in a way that is rather different from how it is understood today), global warming was not regarded as a major issue, or even well understood, by more than a very small group of people until quite late in the 1980s. It needs to be remembered that, when campaigning for incentives for wind power and other renewable energy sources was taking place in the 1970s and until the late 1980s, the justification made for them was hardly ever anything to do with what was then the esoteric and little-known phenomenon of global warming. The justifications for renewable and energy efficiency alternatives were about their role in combating resource depletion and avoiding the problems of nuclear power, and their advantages in combating other types of air pollution, such as emissions associated with acid rain. It is important to bear this in mind, as today renewable energy is promoted heavily on the basis of its being a means of combating climate change. The (partly) changing justification for renewable energy illustrates how 'problems' and 'solutions' are independent of each other.

The point is that the 'scientific facts' as propounded by the Sierra Club were seen through the lens of the conservationist value system that predominated among environmentalists given the social and political landscape of the day (before 1973). This 'landscape', the set of social institutions and value environment, is that which, as Geels (2002) argues, conditions the interaction between technological niches and technological regimes. However, this landscape was to change dramatically with the onset of the 1973 oil crisis. This was triggered by the decision of Arab countries to stop supplying oil to those who were resupplying Israel with arms in the 'Yom Kippur' War. This produced a massive surge in energy prices, and sometimes literal shortages of supplies of gasoline in the USA. However, the Arab states were only able to do this because oil demand was rising more quickly than oil supplies were being developed. Arguably, the politics only triggered what the market would have done a little later.

The point here is that, in the period immediately before the 1973 oil shock, there was an upsurge in environmentalism in general and concern about nuclear power, initially particularly in the USA. Yet, despite the pro-nuclear discourse about rapidly increasing demand for energy, and a counter-discourse from environmentalists that projected increases in energy supply were environmentally unsustainable, the energy crisis did not hit the consciousness of the general public until the winter of 1973. Consequently, before the oil crisis environmentalists could more easily associate themselves with a purely 'conservationist' energy identity than they could after the onset of the oil crisis.

When the energy crisis became a reality for the public at large, and not just a discussion between interest groups concerned with energy policy, the 'landscape', as Geels (2002) puts it, changed. In addition, what interest groups had to say to satisfy the public's altered state of energy consciousness also changed. Perceived interests of different institutions became reconfigured, at least partly by the need to reposition themselves in a changed and now intensely sensitised public conception of energy issues. Hence the selection of scientific and technological facts changed. More attention was paid by environmental groups to the details of energy supply alternatives, and there was immense pressure to offer as convincing a portfolio of responses as possible. Put simply, if people objected to nuclear power stations, then there was pressure to come up with more solutions than merely buying what were now high-cost imports of fossil fuels. Coal was an alternative to oil for electricity, but even that was now much more expensive. Moreover, environmental objections to coal were also considerable.

Lovins's work is remembered for spelling out in technical language what a low-energy (conservation)-based strategy could mean both on the demand and the supply side. Lovins repeated fears about a 'hard' energy path involving large centralised power supplies, especially nuclear power. Besides worries about safety, cost and nuclear waste, 'fission technology also has uniquely socio-political side-effects ... For example discouraging nuclear violence and coercion requires some abrogation of civil liberties ...'.

Instead, Lovins talks about how energy supplies should be based on what is available where people live. He comments:

> Transitional technologies can be built at appropriate scale so that soft technologies can be plugged into the system later. For example if district heating uses hot water on a neighbourhood scale, those tanks can in the long term be heated by neighbourhood solar collectors, wind driven pumps, a factory, a pyrolyzer, a geothermal well or whatever becomes locally available – offering local flexibility that is not possible at today's excessive scale. (Lovins 1977, 48–9)

Anti-nuclear environmentalists were under pressure to come up with some sort of alternative that did as much as it could to offer a supply-side alternative to nuclear power, as well as a strategy based on conservation. This led to renewable energy achieving the status of an alternative energy identity. Ecological progress now became widely identified with an alternative, renewable, energy vision that sought to substitute for conventional energy technologies. Wind turbines became established as an icon of the anti-nuclear or environmental movement. A new discursive (and visual iconic) association between environmentalism and renewable energy conjured into existence a set of renewable energy interest groups that had barely existed beforehand. These interests, in turn, adopted an environmentalist (in the early days often anti-nuclear) discourse that made the earlier statements by Putnam (about conquering nature through use of wind power) seem odd. However, in that case it is again a question of the same interest, but different discourse. Interest and discourse need to be used as an interactive set of tools. One should not be assumed to dominate the other.

As was discussed in Chapter 2, 'mainstream' ecological modernisation analysts have tended to quote Lovins (1977) and his 'soft energy paths' narrative as an example of how amateur idealists cannot supply energy needs, and that ecological change has to be made within mainstream industry. In some ways it is odd to see Lovins criticised alongside so-called anti-growth 'de-modernisers', since Lovins has generally pinned his colours firmly to the mast of enterprise, development and corporate profitability. Indeed, a careful read of his best-known work (Lovins 1977) reveals that, despite his great emphasis on localised, off-grid energy solutions (energy conservation and small renewables), he also espouses support for a 'transitional' use of mainstream, larger-scale energy technologies. However, if we are to understand the role of identity in ecological modernising technologies there is a need to analyse decentralisation's contribution to such ecological energy identities a little more closely.

Decentralisation

In discussing Derrida (1976, 313–16), Howarth argues that 'outside' definitions are necessary for the 'inside' definition of a concept (Howarth 2000, 37). In other words, concepts are defined by 'what they are not' as well as by 'what they are'. This relates to the notion of identity. If being against something is an important point of identity of a group or movement, then there is pressure to construct an alternative identity by emphasising difference from what is seen as being 'the other' – in this case nuclear power. Hence the anti-nuclear energy alternative discourses tended to produce a sort of reverse mirror image of nuclear power. Nuclear power was characterised as being big, centralised, state-directed, remote from the consumer and citizen, elitist and involving (as a discursive justification for nuclear power) rapid growth in energy demand. The alternative was described as small-scale, decentralised, close to the people, supplying local needs and emphasising conservation rather than production of energy. This counter-identity was not just a reaction to nuclear power – perhaps nuclear power was seen to be the apotheosis of an industrialised system that was ecologically unsustainable.

At first glance it might appear that renewable energy advocates have moved away from the early 'decentralised' notions associated with Lovins and Schumacher. However, things are more complex, and this can be related to confusion about different meanings of the term 'decentralisation'. When it comes to energy, perhaps five definitions can be usefully pinpointed: political, ecological, engineering, economic and technological. The term decentralisation is an example of a 'boundary object', a term used by Star and Griesemer (1989) to describe terms that are used with different meanings in different disciplines, although with a residue of common overlap.

In *political* terms it is about local people having control over environmental resources. One widely respected political science text says that a 'distinctive ecological argument for political decentralisation ... holds that policy decisions made at the level of the local community should be more sensitive to the environment' (Carter 2001, 56). In the case of wind power, analysts sympathetic to the Danish wind power programme argue that the system of locally based ownership (by cooperatives or farmers) boosted local acceptance of the wind turbines. Frede

Hvelplund, one of the key advocates of the early Danish renewable energy programme, has commented: 'People seem to like wind turbines, when they own them, and are not annoyed by the noise and visual inconveniences; especially when receiving a fair compensation. However with a system of distant utility or shareholder ownership, the local inhabitants are only getting the disadvantages without the compensation. This is seen as unjust and results in increasing local political resistance against wind power. It is as simple as that' (Hvelplund 2005, 237).

The *ecological* definition is expressed by an Executive Director of Greenpeace in a foreword to a 2005 report by Greenpeace UK entitled 'Decentralising Power': 'In a decentralised energy (DE) system, electricity would be generated close to or at the point of use' (Greenpeace 2005). A more technical definition that amounts to more or less the same thing is that it is 'connected directly to the distribution network or on the customer side of the meter. This means that the electric energy only needs to be transported over short distances, thus reducing transportation losses. The most common technical term is "distributed generation" (DG)' (Ackermann et al. 2001; Karger and Hennings 2009, 584). However, by this definition the larger wind farms now being built in places ranging from Texas to China and off the north European coast would not be counted as 'decentralised', as they will be connected to the high-voltage transmission system before anything is distributed. That is the case, for example, with large British and Belgian planned offshore wind farms, which will be connected to 400 Kv transmission lines (Econnect 2008; SenterNovem 2005).

The notion of using energy that is produced near its point of use also appears to be the meaning employed by Scheer (2007), an SDP politician who is regarded as a key advocate of the feed-in tariff policy in Germany. He recounts the classic end-of-nineteenth-century contest between Westinghouse and Edison: Westinghouse's system of long-distance transport of electricity through AC lines won the day over Edison's vision of local generation with DC distribution. Scheer argues that renewable energy could provide a more sustainable solution compared with the coal that Edison's model would have employed (Scheer 2007, 61–5). Among the energy sources that fit the possibility for providing such 'close to point of use' supply are solar, wind, small hydro, heat pumps, combined heat and power (CHP), biomass and anaerobic digestion (biogas) (Tech-wise 2002; Wolfe 2008). This point-of-use concept also ties in with Lovins's soft energy paths notions (Lovins 1977) and with early 1970s theory about local sustainability. If, as the theory goes, resources are obtained locally, there can be much greater aware-ness of whether the carrying capacity of the land is being exceeded than if the resources are supplied from far away.

This *ecological* concept of decentralisation is often also linked to *political* decentralisation through pointing out that local production of energy can be organised by local actors, thus acting to 'democra-tise energy... By enabling local action and empowering individuals and communities as producers, decentralisation has the potential to bring about a massive cultural change in our attitude to and use of energy' (Greenpeace 2005, 5).

The *engineering* definition of decentralisation involves the lack of con-trollability of the energy source. Variable sources, such as wind and solar, generate energy according to the flows of natural forces rather than the input of fuels as is the case with fossil and nuclear energy sources (interview with Xiao-Ping Zhang, 24 February 2009). Thus for electrical engineers it does not matter how large or small a wind farm is, or how local or remote from the point of consumption; it is still 'decentralised'.

The *economic* definition of decentralisation involves trading deci-sions being made by non-centralised actors. Hence the replacement of a system of state decision-making by the market is usually the effective meaning of the term. Economic decentralisation can also refer to 'fis-cal' decision-making being given to lower levels of an administrative hierarchy (World Bank 2009). It could also mean the transfer of deci-sions made by formerly private bodies such as retail electricity monop-olies to liberalised markets, as has happened across the world since the 1990s. As we shall see in Chapter 5, the notions of market-based or 'command and control' styles of decision-making in renewable energy financing are a source of considerable controversy, perhaps for rather confused reasoning, as I shall explain. However, advocacy of 'market-based' systems of environmental interventions in energy, such as 'cap and trade' schemes, has become popular in the West since the end of the 1980s.

A fifth, *technological*, notion of decentralisation is provided through the work of Schumacher (1973, 159–77), especially since he is often quoted as a source of thought about 'decentralisation'. He talked about 'intermediate' technology (IT) being 'appropriate' to human scale development. There is an implication of smallness here, although Schumacher said this was about its affordability for local people. His focus was particularly on the developing world, and IT also implies that technology is influenced by the needs and actions of the people on the ground rather than some large remote institution. Tidal stream devices were first conceptualised as being 'IT' technology for use in the developing world, although more recently the technology is at a demonstration stage in the industrialised world (interview with Peter Fraenkel, 17 July 2008). The IT movement gave support to wind power in the 1970s, and was influential among thinkers in California.

Perhaps the most curious thing about Lovins and others in the 'small is beautiful' (IT) discourse is that, while they were indeed the talkers, not the 'doers', of the new renewable energy industry (although, as will be seen in Chapter 4, very important talkers), the doers themselves can hardly be classified as being part of the mainstream energy industry. The doers were a highly effective group of largely Danish innovators who virtually invented two new industries at low cost and in only a few years, in contrast to the inefficient, expensive and ineffective efforts of the mainstream electricity and engineering industries (Toke 2011).

Invention of a new technology

We can see two things from this discussion. First is that on its own decentralisation, though clearly distinct from nuclear power, represented too vague an alternative to be of practical use without some clear ideas on energy supply options. There is something (in decentralisation) of what Laclau (1980) calls an 'empty signifier', a term used to unite opponents of something that in reality does not have much common meaning in practice. Decentralisation is only a partly positive measure to justify maintaining an identity of opposition to nuclear power. A second point is that the identity of renewable energy as a technology is something that needs to be analysed as distinct from the notion of decentralisation. It is an identity that is more precise, which provides the fullness of a conceptual response, at least, even if arguments remain about how practical it is.

Renewable energy sources such as wind power and solar power have an identity today, and this is something that hails from the energy politics and context of the 1970s, as an alternative to conventional nuclear and fossil fuels while at the same time being seen as environmentally sensitive as well as being immune to resource depletion. The very strength of renewable energy identity lies in being able to marry the twin pillars of environmentalism and 'clean' industrial technology. This is the essence of ecological modernisation and provides an illustration of how the concept of 'identity' EM needs to be deployed in order to analyse the development of renewable energy.

The modern invention of wind power, which occurred in Denmark in the 1970s and 1980s, needs to be discussed by reference to some specific traditions and the way that these traditions, based on local reasoning, interacted with new dilemmas to produce outcomes (Bevir and Rhodes 2003). The argument produced here is that decentralisation did not so much provide an identity for the new technology of wind power (and other renewable energy technologies) as describe part of a pragmatic context and market in which new technologies were born. Indeed, the new technologies, such as wind power and some others such as farm-based biogas, were themselves descendants of existing traditions. Denmark, among other countries (including the USA), had had a tradition, since the beginning of the twentieth century, of off-grid use of wind turbines to provide electricity; similarly in the case of biogas, which had a long pedigree of use in India and also on farms in Germany in the 1950s (interview with Arthur Wellinger, 6 January 2010). By contrast, the relatively late development of wave power and tidal stream power (mainly in the UK) could be attributed to a lack of a tradition in these technologies. Solar photovoltaics, which we shall mention later, came about as a new industry, radar, coincidentally provided an answer to an old question.

Denmark's wind power traditions are significant in that the horizontal-axis, three-bladed, upwind design that was developed in Denmark has come to be the standard for the world today. This tradition was ensconced in the 'Gedser' prototype machine built by Johannes Juul in Denmark in the 1950s (Hau 2006). However, there are plausible arguments that vertical-axis machines[2] have advantages, especially for applications such as offshore wind power (interviews with Derek Taylor, 18 March 2009 and interview with Andy Bowes-Lyon and Annie Hairsine, 24 June 2009). However, the cost of developing such an alternative deters the adoption of new technology. Technological path dependence can be seen in operation even in what is regarded as a relatively new modern reincarnation of a technology such as wind power.

What can be called the dominant Danish technical tradition of wind power cannot, of course, be separated out from a Danish social tradition of engagement with the technology that was strongly bound up with the importance of the rural agricultural sector even in the 1970s in Denmark. This tradition was uniquely concerned with two notions that had great significance for the development of wind power. One was a tradition of local self-sufficiency and another was the notion of sharing knowledge. The social movements that have supported wind power and the theoretical connections between social movements and ecological modernisation are discussed at length in Toke (2011). The early development of wind power is an example of a 'technology product movement', a notion coined by Hess (2007). According to Preben Maegaard, who became the Chairman of the Danish Renewable Energy Association in the late 1970s:

> Denmark was the country most influenced by the energy crisis and almost 100% all supplies – transport, power production, heating and so on – was based on oil and this oil came from Arabian countries. You could not find another country that was so dependent on oil and therefore the reaction was very strong in the general public and the establishment, its response to it, that was to say 'now we have to turn to atomic energy'…This was what the establishment did, but there was a reaction in the population saying that that would create a new dependency. And there came some movements against that. And the question was how to get involved in that and that was by saying we have natural resources we can use – the sun, the wind and the biomass. (Interview with Preben Maegaard, 18 April 2009)

We can see here an implicit reference to a notion of national self-sufficiency, which dates back to wars in the nineteenth century and also a rural education movement led by the evangelist Gundtvig (Allchin 1998). A related tradition is that of sharing knowledge in the Danish agricultural sector, and this practice was embedded in an 1885 law that prevented agricultural technology being patented (interview with Preben Maegaard, 18 April 2009).

We can also see how these traditions had to interact with a perceived dilemma – that of meeting the challenge of the 1970s oil crises without the nuclear power to which many were opposed. The result was a grass roots and wider social movement in which ordinary people, craftsmen and also technical and academic specialists gave their unremunerated spare time to develop new technologies of wind power and biogas (Asmus 2001; Eyerman and Jamison 1991; Garud and Karnoe 2001; 2003; Heymann 1998; Karnoe 1990; Kemp et al. 1998; Nielsen 2001; Olesen 1998; Toke 2011; van Est 1999). The movement constituted an alternative paradigm based on a set of non-material values involving a wish to avoid the ecological pitfalls of nuclear energy. However, this ecological rationality was also justified in rational terms, with a belief (contrary to the views of the electricity utilities) that wind power could generate power at costs that would be competitive with power from fossil fuel and nuclear sources.

Many farmers and blacksmiths started making designs for windmills, the Juul prototype being influential. The first grid-connected wind power generation occurred in 1975. The construction of the first grid-connected machine is attributed to Christian Riisager, although, more correctly, his machine was the first to be given consistent coverage in the press. Riisager took orders for numerous machines, using the same plans (relying heavily on Juul's ideas), although each machine was made at a different blacksmith's, meaning that all the machines were slightly different. The initial market was to sell machines to enthusiastic farmers to supply their own needs and send the surplus into the grid. The market was a very open one, with considerable information flows. Production data were shared and there were meetings organised at which turbine owners, activists, farmers, technicians and manufacturers attended to compare machine performance and optimisation of techniques.

Around twenty wind turbine manufacturers had been established by the end of the 1970s, although most of these were short-lived. The machines were small ones by today's standards. A particularly important development was the construction of what was then a giant windmill at the Tvind High School. This institution was established in the Grundtvigian tradition and the teachers used money for their wages to build the machine. They were assisted by a range of technicians and engineers giving their advice free and working in their spare time. Among the helpers giving their efforts for free were technicians from the Danish Atomic Research Station at Risoe (in their spare time), and also the pioneer wind turbine developer (in Germany in the 1930s), Professor Hutter from the University of Stuttgart. Local engineering firms also gave considerable assistance, for example, in helping design mouldings for turbine blades (Interview with Joep Nagel, Tvind Folk High School, 10 May 2009).

As one of the Tvind High School organisers says:

> the whole philosophy was to develop wind power and to make it viable and to support the popular movement that was taking shape in that market to develop wind power. You have to realise that at the time it (the size of the proposed machine at Tvind) was completely ridiculous and the largest windmill that was actually working was maybe 20 kilowatts. A lot of people, and I think this very, very crucial to understand it, the reason why we succeeded was that we were an open project. All our reserves were available for everybody, especially the wings (turbine blades), people could come and have all the

drawings, could have all the forms for the wings and the windmill as such. And I think at that time we built the large windmill, the 2 megawatt one, but we also built smaller windmills and the wings of that were used on many other windmills built in Denmark and has been actually the starting point for many of the commercial producers to produce wings for windmills. (Interview with Joep Nagel 10 May 2009)

As we have seen in Chapter 2, there is a tendency for mainstream EM theorists to decry attempts to develop markets for ecological products outside mainstream industry, seeing it as uncompetitive and incompetent. Yet in fact the free flow of information and the tremendous desire to share innovation ideas proved ideal for developing the technology. It may not be typical of mature industries, and this type of activity has long since disappeared from the wind industry, but it must nevertheless be considered as an important part of the EM process in the case of wind power and other renewable energy technologies. It is essential, and the technology would not have been developed, at least not so quickly, without this beginning as a popular movement. It was a movement that, ironically, became so effective at designing a new technological pathway for energy precisely because it was inspired by non-material values and idealism rather than financial gain in the short term. This was done because, as implied earlier, it coincided with a Danish social tradition of how rural industry should work in solidarity and cooperation in times of national crisis.

the development did not cost any money because we worked without money in that group. There people who could do design work and engineering work and there was one in the group who could make the electric controls so they were professionals in their fields...This was interesting and important and we wanted to develop wind turbines to make local production and also so people could cover their own energy needs and so on...idealism is a strange thing, you see, people do many strange things in life and why should they not make fun?...If the opposite of idealism is work then work is punishment in your life. Of course, idealism is different from that but it's a very special and difficult to discuss because why do people do sport, why do they paint, why do they do many strange things, what could come out of that? This grew out of the fixation and I think it was a natural attitude for people growing up in the Grundtvig culture. (Interview with Preben Maegaard, 18 April 2009)

The popular movement pressed the electricity utilities to pay money for the electricity supplied to the distribution system at premium rates. At first ad hoc arrangements appeared, negotiated with individual utilities. A grant system was introduced by the Danish Government in 1979 to support the new technology of wind power. By 1985 a coherent system of incentives for wind power had emerged. It consisted in part in paying investors in wind power the taxes they paid on their electricity consumption emerged. This incentivised members of wind power cooperatives. Also a standard national price was payable to wind turbine operators for each unit of electricity that was generated. This was the first instance of what became known later as a feed-in tariff system. The main difference between this and later versions, the best known version of feed-in tariffs being in Germany, is that in Denmark the Government repaid part of the effective subsidy given by the utilities to the consumers. In other feed-in tariff systems the subsidies are reclaimed from levies on electricity bills paid by the consumers. In fact the reliance on Government revenue (as opposed to electricity sales revenue) later (in 2001) proved to be the 'Achilles heel' of the system.

The market then developed for slightly larger machines to supply cooperatives. Wind Power cooperatives in Denmark evolved in the context of a very Danish tradition favouring cooperative organisation. Part of this is no doubt connected to the fact that the country continued to have an important agricultural industry very late compared to other EU states, and also partly due to the importance of cooperatives to the way people organise themselves more generally. Cooperative combined heat and power projects have been an important part of the country's response to the oil crises of the 1970s, and cooperatives also feature strongly in other parts of Danish life, ranging from housing to the organisation of locally controlled schools. At the end of the 1970s wind power cooperatives began to start up, fired partly by an idealistic anti-nuclear response to the energy crisis. These were organised on the basis of the windmills being purchased entirely from equity raised by local people. In the 1980s the trend took off, and by 1990 there were some 150 MW worth of cooperative wind power projects compared with around 60 MW owned by private individuals (mostly farmers) and 40 MW put up by the utilities (Global Watch Mission 2004, 21). This sounds small by today's standards, but even at the end of the 1980s the optimum size of a wind turbine was still only around 200 KW. Vestas began its march towards being a multinational company with the successful marketing of a 55 LW machine in the early 1980s.

Of course, even the 250 MW implemented in Denmark by the beginning of 1990 pales into relative insignificance compared with the 1400 MW of wind power that was installed in California mostly between 1982 and 1987 (Starrs 1988, 109). As is discussed in Chapter 4, it was the 'wind rush' in California that turned a near cottage industry into a more conventional industry. Nevertheless, cooperatives, from the late 1970s until the early 1980s, provided the vital development of a market that led to larger machines and the modern industry.

It is also necessary at this point to stress again that, because it was only proto-industrial, and heavily laden with idealistic motivations and often amateur in endeavour, this period of activity is not easily classified by the mainstream EM analysis as important. However, without the initial (re)invention of the technology in modern grid-connected form by idealistic groups of farmers, activists and technicians, the more mature industry would not have occurred. Mainstream EM seems to have identified industry with conventional, established industrial groups. However, industry and technology can, in theory, be developed by anybody, and the electricity generation industry in its modern form is largely concerned with grid-connected industry. The Danish wind power revolution was clearly about the development of a grid-connected industry, even though its social origin arose from decentralised political and engineering activity. To return to the central theme of this book, the emergence of this revolution was concerned with a new technological identity resulting from a reconstruction of an old technology in modern form in the context of a new set of ecological and societal priorities.

Wind turbine installation took off in the 1990s in Denmark. However, as the 1990s wore on, the installation of wind turbines by farmers outpaced the growth of cooperatives. Each farmer was allowed to install one machine on his or her land, and this proved to be so profitable that there was an explosion in the rate of construction. By the time a right-wing political coalition took over in Denmark in late 2001, and effectively ended the onshore wind power development programme, sufficient wind power was in place or planned to provide nearly 20 per cent of Danish electricity from wind power. That proportion has increased a little because of a continued offshore programme. Nevertheless, despite this growth, the German wind power market took off in the 1990s, and over 6000 MW had been installed in Germany by 2000, more than double the amount installed in Denmark. Germany thus became the biggest market for the wind generator manufacturing industry, which was in a process of maturation. Danish companies such as Vestas and BONUS led the world market, although BONUS was later taken over by the German company Siemens.

The onshore wind programme in Denmark came to a halt after the elections of November 2001. The wind programme had already become increasingly controversial because of the perceived cost to taxpayers. Since then other feed-in tariff systems have been funded by levies on all electricity consumers, and thus have been less prone to competition with competing demands for government spending. The 2001 election was the first time a solidly right-wing coalition had come to power in Denmark in several decades. The right had been generally hostile to the subsidies for wind power, but had been unable to act on such feelings, since, even when it was in power, it had been prevented from stopping the subsidies because coalition centre parties supported the feed-in tariff system. However, since 2001 the right-wing coalition has had an overall majority in the Folketing (Parliament), and the Government has only funded a modest offshore wind programme (Ryland 2010).

One characteristic that was not included in the heuristic used to measure identity EM in Chapter 2, but which might arguably be added, is the extent to which the wind generator manufacturers are independent of the energy industry. I decided not to include this as a characteristic of identity EM because it may be that wind generator manufacturers are owned by, say, engineering companies that are not first and foremost holders of assets in the energy industry. Nevertheless, it is possible to observe that, while the largest company, Vestas (Denmark), remains independent, as well as Enercon (Germany), Siemens and General Electric are now key players in wind power manufacturing, having acquired leading wind generator manufacturers. Siemens acquired Bonus and General Electric bought up Zond. Iberdrola, which is focused on the electricity industry, retains a close relationship with Gamesa. In China three companies, Sinovel, Goldwind and Dongfang Electric Corporation (DEC), lead the Chinese wind generator market, and are likely to make increasing inroads into Western markets in the future. Of these three companies, Sinovel and Goldwind specialise in wind turbine manufacturing while DEC's wind operations are part of a wider electricity generation portfolio. Sinovel claims to be the first Chinese company to be 'independently developing' wind turbines in China (Sinovel 2010).

Invention of other new technologies

The bulk of my attention has been focused on wind power's emergence as a 'new' renewable technology partly because it has become the predominant 'new' renewable energy technology today. A further reason for a focus is that wind technology has become an icon, an easy point of identification, for the energy 'alternative' and environmental approaches. This is sometimes ironic, given the strong opposition that can attend the proposal of wind farms in some parts of the world, for example in the Bay of California and the hills of Wales. Nevertheless, from the point of view of analysing renewable energy identity, as well as the need to focus on what is the largest provider of capacity among the 'new' renewable energy technologies, a principal focus on wind power is justified.

Although this chapter has gone into some depth about the development of wind power's identity, there are insufficient resources to do the same for the other renewable energy technologies. However, it is pertinent to spend a little time giving an outline of how some other renewable energy technologies may fit in, or not, with the patterns and analysis observed with wind power above.

In fact, another technology was virtually invented in its modern form in Denmark in the 1970s. This is biogas, and it is worth spending a little time on this since it shares part of the 'social movement' path to innovation that is characteristic of the development of wind power in Denmark. Like wind power, this had a previous tradition. As in the case of wind power, the oil crisis of 1973 sparked an interest in modernising the technology. Again, as with wind power, information about optimising the processes was freely exchanged in a variety of meetings at which a wide variety of farmers, craftsmen and activists took part.

Preben Maegaard, as Chairman of the Danish Renewable Energy Association, took a keen interest in the development of biogas technology in the 1980s. This technology, like wind power, had a social tradition. Maegaard recruited expertise from India using a grant obtained from UNESCO, although this expertise was decried by establishment academics in Denmark:

> there was a leader in the biogas programme from the Institute of Technology...and he said 'To make biogas is just too specialist', 'used a primitive design', he actually did say so, he humiliated people by saying 'With such support from India will never work. We must have some professionals involved.' In the end, after some years, it was only these Indian influences that worked well on the so-called professional plants, disappeared again. And that is the history. (Interview with Preben Maegaard, 18 April 2009)

A key difference from the case of wind power is that, while the nascent wind power industry received support from academics (later professors) such as Frede Hvelplud and Niels Meyer, the biogas experimenters lacked such support. Raven (2004), discussing developments in The Netherlands, reports how the emphasis of the professional technical establishment was on 'centralised' biogas plant, rather than farm-based equipment, and this held back technical developments. However, the 'amateurs' persevered, supported by the Danish Folk Centre. In fact, in the case of biogas, much of the inspiration and, later, wider deployment of the technology came from Germany. This included efforts to spread knowledge about the technology by idealist groups such as 'Bundschuh' group that was based in Bavaria. This was one of various groups and individuals spreading information about the technology. Arthur Wellinger, who much later became President of the European Biogas Association, organised tours of biogas plant for the purposes of showing farmers how the technology works (interview with Arthur Wellinger, 6 January 2010). The spread of knowledge through idealistic enterprise (not directly linked to any profit-making activities) formed the basis upon which the more recent growth of the biogas industry in Germany has been based.

In the 1990s some larger-scaled plant, sometimes adapted to industrial waste-producing companies, emerged, the size of plant increasing on an incremental basis much as in the case of wind power. Sometimes the plant involved combined heat and power as an alternative to electricity generation. These plants were commissioned as a result of incentive programmes in Denmark and Germany. However, biogas as a technology began to take off after the year 2000, when the German feed-in tariff was amended to give high rates for electricity produced from biogas as opposed to receiving the same (lower) rates for wind power and hydro power under arrangements that had been in place since the inception of the feed-in tariff in 1991. By the end of 2007 there was 1280 MW of biogas electricity-generating plant in place (Burgermeister 2008). By this time biogas was supplying around 1.5 per cent of German electricity, with this proportion rapidly increasing (Cordsen 2009). Projects are now being organised that feed the biogas directly into the natural gas grid rather than merely acting as a feedstock for electricity production.

The main focus in this chapter has been to look at the emergence of the leading new renewable energy source, wind power, and, to a lesser extent, biogas, as examples of how 'identity' ecological modernisation processes can often involve the wider public in supporting, and, in some key moments, actually participating in the technical as well as social (re)construction of renewable energy technologies. Although there is not enough space to do a full review of the various renewable energy technologies, it is useful to briefly sketch out the position of other technologies in this field.

Some technologies, such as geothermal energy and biomass-burning technologies, have been heavily based on existing steam turbine technologies, with the difference being that the heat source is different. In the process, of course, there has been some optimisation; for instance, in recent years small biomass burners have been adapted from fossil fuel-burning ones to deal with the specificities of burning biomass for both electricity and heat-only production. Also, there have been some attacks on the renewable identity of some biomass sources of energy, especially biofuels for transport purposes. It is not the objective of this work to analyse such debates, although it is relevant to note that the continued likelihood of such energy sources retaining incentives and regulatory advantages over conventional fuels depends largely on the maintenance of a positive public identity.

Other, entirely new, renewable energy generation technologies have been or are being developed through, compared with wind power and biogas, relatively conventional industrial innovation pathways. Solar pv in particular was invented in Bell Industrial Laboratories in the early 1950s as a consequence of new insights in the field of using silicon in transistors, this being a spin-off from the use of silicon for its semiconductor qualities in radar in World War Two (Green 1993, 338; Perlin 1999, 25–34). However, as will be discussed later, for example in Chapter 6, in the section on Germany, the implementation of solar pv has been achieved partly though broad-based public support and enthusiasm, and its progress is largely bound up with very positive public identification with the technology. According to Professor Martin Green:

> The technical focus these days is mainly on refining manufacturing processes and also there is a push to differentiate the product. However, the basic technology is much the same as it was in the 1970s except that manufacturers are aiming for cost savings such as by using bigger wafers and higher throughputs with the aim of achieving grid parity with other electricity sources at point of use. (Interview with Martin Green, 3 August 2009).

Some variants of solar power, including different solar thermal–electric techniques (solar trough, power tower) and centralised solar power (CSP) fit into a more large-scale industrial identity compared with small-scale solar pv. As will be discussed in Chapter 5 on the USA, there is evidence that such industrial technologies are preferred by major electricity suppliers to wind power for a given price.

Marine-based technologies, wave power and tidal stream technologies, have also been developed through conventional industrial innovation patterns. Yet their progress so far has been slow. This may often be ascribed to difficulties in overcoming technical obstacles. A more sociological approach would point to a lack of traditions in such technologies, connected to the somewhat obvious fact that people do not live in the sea. If mermaids were real, no doubt they would have started wave power and we would be a lot further advanced in developing this technology than we are at the moment.

Conflicting identities

This book focuses on the positive technological identification with renewables – in order to rectify a lacuna in the analysis and use of ecological modernisation theory in which, as seen in Chapter 2, technology is seen as a matter that is mostly the creature of conventional industry (in the mainstream version) or (often in the 'radical' version) as a 'problem' facing civil society. Indeed, the politics and planning of renewable energy have usually been analysed in terms of its problematic characteristics involving planning controversy (e.g. Haggett and Toke 2006; Toke 2002a; 2005; Toke et al. 2008). Nevertheless, just because this work follows a wish to make up the balance by analysing the ways in which publics engage with renewable energy as a technological identity that is distinct from conventional energies, this does not mean that this work should totally ignore the planning controversy. This also involves issues of identity, or, perhaps more accurately, issues of conflicting identities. The aim here is to sketch out a response that outlines what such an 'identity' analysis should be. We need to begin by summarising the debate on the subject of wind power planning controversy.

Wind power planning controversy in the UK, for example, has been conceived as involving a 'gap' between what is measured (by opinion surveys) to be overwhelming public support for 'clean' renewable energies including wind power and the frequent controversy surrounding planning applications for wind power at a local level (Bell et al., 2005). The most popularised explanation for the 'gap', at least in terms of the narratives of wind farm developers, is 'NIMBYism'. One definition of 'NIMBYism' is when people oppose projects simply because they are close to them, that is, as opposed to disliking them as a matter of principle. However, this explanation is criticised by Wolsink (2000) using some rigorously assembled quantitative analysis of opinion surveys which concludes that 'NIMBYism' is a myth. He says that local opposition to wind power can be principally explained by the existence of generalised opposition to wind farms, since few people surveyed actually say that they are in favour of wind farms so long as they are not near them.

There is a problem here, which has been identified by van der Horst (2007). What people may say in answer to opinion survey questions (for example about their motivations for opposing developments) cannot necessarily be regarded as being a positivist (absolute, in all contexts) expression of their interests. Rather, a response to a survey question could be regarded as a discourse that should be seen in a particular context, that is, to support a position by claiming to support fair debate, espouse a defence of universal principles (e.g. about landscape values) and to use such claims to avoid admitting to local (and therefore 'selfish') concerns. Certainly Devine-Wright argues that the term is confusing: 'the NIMBY concept unhelpfully muddles whether opposition should be conceived as a belief or attitude towards a development, a behavioural response taken by individuals or the collective actions of organised groups' (Devine-Wright 2009, 431).

The issue that van der Horst (2007) raises is whether the problem can be solved by quantitative analysis of survey responses. Wolsink's (2000) attack on the notion of 'NIMBYism' assumes that it has a precise definition and that particular subjects can be said to have a measurable position on the subject. However, one does not have to be a convinced postmodernist to understand that the truth of this issue is difficult to decide outside discourses used in particular contexts. Indeed, Wolsink's own data and arguments demonstrate that the proportion of opponents of a particular wind proposal varies across time, with the greatest opposition occurring, according to Wolsink (2007, 2696), during the planning application phase. Yet this is in reaction to a particular proposal in a particular place and time, and there are philosophical problems with assuming that the actors involved in this debate are solely influenced by universal, as opposed to case-specific, factors. However much those surveyed may say they are not 'Nimbys', the issue of their identities cannot be settled outside the narratives being used on one or other side of the planning dispute. The developers will call the objectors 'Nimbys'; the objectors will deny this. As is argued later, rather than trying to settle whether people should or should not be described as 'Nimbys', it is plausible simply to analyse the conflicts in identity that may occur in a given planning controversy.

Not content with pure analysis, Wolsink (2007) looks towards a normative agenda to discuss ways of overcoming this opposition to wind farms. He cites surveys that report popular support for deliberative methods of consultation over wind farm planning applications as support for this. Indeed, such a narrative fits in well with the type of deliberative institutional agenda favoured by Hajer (1995).

The 'deliberative' approach to ecological modernisation, which has been criticised in Chapter 2 as failing to sufficiently tackle issues of technological identity, may have limited utility when it comes to normative dimensions of wind power planning issues. Indeed, Devine-Wright comments that: 'it cannot be assumed that deliberative public engagement in renewable energy developments will secure public acceptance. In fact, it may cause the opposite, providing a means for local people to collectively organise and communicate their concerns within an interactive process' (Devine-Wright 2007, 9). There is something of a contradiction in the 'deliberative' argument. If opposition to wind farms is largely associated with generalised opposition to wind power, with localised opponents incentivised to campaign because their own 'local end' of landscape protection interest is challenged, what exactly is the benefit of more deliberation? Since the implementation of the Aarhus Convention (UNECE 1998) in the EU there is a statutory necessity to consult all interested parties thoroughly in the planning process anyway, so what can more deliberation achieve (beyond this), except perhaps to slow planning applications down even further?

Wolsink (2007) argues that people want an increase in deliberative practices, but then those who are sceptical or opposed to the proposals certainly would argue for more discussion, consultations, the production of more reports on environmental impact issues (usually grossly unread) and so on. Yet this will act as a barrier to wind farm development and therefore help the opponents, not achieve a 'balanced' outcome, whatever that may be. Such arguments for more deliberation should not be confused with incentives that might make wind farms more attractive, such as job creation through local manufacture of parts, giving money to the locality through grants, taxes or even direct payments to homes in the immediate vicinity. These are more practical, and perhaps useful, ideas.

Confusion occurs in the minds of environmentalist analysts when they assume that more deliberation is a general cure-all for environmental problems. Support for deliberation is especially strong among environmentalists, since it helps those who are sceptical of top-down industrial plans to impose technology on doubtful citizens. However, just because this helps environmentalists, for instance in opposing plans for large centralised power stations, a fuzziness of logic leads them to believe that more deliberation will lead citizens to understand the positive logic of technological plans (such as wind farms) to which the environmentalists are more sympathetic. They seem to assume that the meeting of demands for more deliberation will lead to more acceptance of wind farms. Unquestionably there should be a citizen's right to a reasonable amount of deliberation, as argued under the Aarhus Convention. However, as Bohn and Lant (2009) observe in the USA (and as discussed in Chapter 5), when there are fewer arrangements for consultation and deliberation over wind farms this tends to be correlated with less opposition and greater planning acceptance of wind farms. The point is that demands for more deliberation tend to arise when there is more opposition to wind farms. When wind farms are uncontroversial, there are few demands for more rigorous planning procedures and more deliberation. Of course, it may be that in some cases the standards outlined in the Aarhus Convention (and implied by EU Directives) have not yet been achieved, but in places such as the UK and The Netherlands there is now plenty of opportunity for consultation on wind power planning matters.

In Spain there has been a tradition, until recently at least, of less consultation at a local level, and analysts such as Zografos and Martinez-Alier (2009) perceive a deficit in deliberative arrangements. This may be the case, but it would be wrong to assume that improvements in 'procedural' arrangements will reduce conflict over wind farm siting. Indeed, any pathology of generally lower facilities for local consultation in Spain is more likely associated with a tradition of lower levels of local opposition to wind farm proposals in most parts of Spain compared with, say, the UK or The Netherlands. Similar comparisons can be made between different US states, with much lower levels of opposition to wind farm plans in most areas of Texas compared with, say, the Bay area of California. In Texas there are usually many fewer planning hurdles for wind farms compared with California.

Increasing opportunities for deliberation will not usually increase the possibilities for wind farm acceptance, however necessary it may be to improve deliberative processes to remove deficits in citizens' rights. Indeed, if every demand for increased deliberation by wind farm opponents is granted, the effect will be to delay any opportunity for construction indefinitely. The point of deliberative processes, according to their advocates (Renn et al. 1995), is to achieve as near as is possible to a free, even-handed discourse. Some would interpret this, in the wind farm case, as meaning that a wind farm developer would have, for example, to allow a discussion about where the wind farm should be sited in an area, or should allow a positive planning decision to be open for challenge for up to 4 years after consent has been granted (this latter notion is mooted in France). Yet all of this would certainly not be viewed as 'fair' by a developer. An open-ended debate about whether an area should have a wind farm may not only close off the practicable (economic) options that may be available to a developer, but also allow diehard opponents of wind farms to paint the worst possible scenario for wind farm development. The ability for post hoc challenges to take place means that the project could be delayed for several years until the challenge period had elapsed. The point here is that what is meant by free and fair deliberation in renewable energy planning cases is open to a lot of interpretation.

If wind farm planning controversies are not satisfactorily analysed or normatively solved by appeal to deliberative theory, what analysis can we use? One path is to use notions of identity. Devine-Wright (2009) looks to psychological theory by referring to the strength of 'place identity' (how people identify with their localities) as being part of an explanation for how local people may object strenuously to wind farm planning applications. This involves a local identity that conflicts with the positive technological public identification associated with wind power. However, notions of place identity may not connect firmly enough with the generalised criticism of the impact of wind farms on landscape upheld by national organisations such as the Council for the Protection of Rural Wales or the UK Ramblers Association. Put another way, conceptions of local place identity can themselves be associated with nationally or regionally based traditions of concern for landscapes. Is this, therefore, to be described as a local identity or part of a national tradition of landscape aesthetics? Other national traditions may be different. After all, in some countries, such as Spain, landscape protection is not sufficiently appreciated to merit the existence of national landscape protection organisations.

What we need is to deploy a framework of 'conflicting identities' to understand wind power planning controversies across a range of cases. Mouffe (1995) talks about how the search for consensus in democracies may be undermined by the existence of different identities which are sometimes different to reconcile. Analysts tend often to search for, or assume, a single universalist identity for actors in a particular case. Yet it may be more useful to analyse problems as if there were several identities in operation at different times and in different places. This is well illustrated by the material already discussed in this section; for example, some countries, such as the UK, are marked by traditions of landscape protection, whereas others, such as Spain, are less so (Toke et al. 2008). In other cases attitudes change over time. For example, Breukers discusses how in the 1990s there was much support for wind farms in Nord-Rhine Westfalia, couched in terms of support for the technology's role in combating air pollution problems, but that in more recent years this support has waned in favour of concerns about the local impacts of wind farms (Toke et al. 2008).

Hence, we can list several conflicting identities. First, we have the eco-technical identity of support for wind farms (or the renewable energy source under discussion) as a measure to supply clean energy. Second, there is what Devine-Wright (2009) calls the 'place identity'. Third, there is, in some countries at least, what could be called a 'landscape identity', as articulated by national organisations. It is also possible to identify a fourth 'developmental' identity, which sees, often from a regional point of view, the wind farms as an opportunity for creation of jobs and for taxation. This dimension is discussed in Chapter 6 in the contexts of Spain and China. It may be by discussing the role and interaction of such identities that we can analyse planning issues surrounding wind farms and, moreover, be able to understand the outcomes that transpire.

Conclusion – from local supply to supergrid?

We can see that, in discussing the eco-technical identity associated with renewable energy, a paradox emerges in that, while a doctrine of providing local needs was used in the past to help renewables build a foothold as a nascent industry, today the talk among many radical environmentalists is about the need to transmit renewables across distances. Both Greenpeace and Friends of the Earth are vigorously promoting the development of international interconnectors to help renewable energy be absorbed in the grid. Environmental NGOs seem to be taking their cue from the mainstream renewable energy industry, which is working with the EU to develop more comprehensive international interconnector systems, or, in the USA, backing regulatory initiatives to ensure building of transmission infrastructure that can carry renewable power across long distances. Greenpeace has funded a series of reports variously favouring large-scale offshore wind farms for the UK and investment in long-distance electricity interconnectors to service multi-nation efforts to develop offshore wind farms (AEA Technology 2002, 3E 2008).

Governments and their energy advisors have accepted renewable energy technologies such as wind power as being an important part of the strategy to counter global warming and fossil fuel depletion, although, in contrast to the anti-nuclear activists whose actions developed modern wind power, the renewable contribution is seen as complementary to that of nuclear power rather than in opposition to it (MacKay 2009). In addition to this, some large energy companies have thrown their weight behind the notion of Europe deriving much of its power needs from giant installations of concentrated solar power in the North African desert, the so-called 'DESERTEC' concept (Clean Techies 2009). Some activists associated with green NGOs, such as Hermann Scheer and Eurosolar, have taken issue with this vision, preferring to rely on more local, or at least country-specific, sources (Scheer 2007; 2008). Is it that the identity of renewable energy has shifted, even in the minds of the environmental campaigners themselves?

Certainly there is some scepticism from Scheer and others towards plans to build large amounts of solar thermal electricity generation in North Africa. This argument relates to the issue of public identification with renewable energy sources, since it is doubtful whether energy consumers in one country will identify with developments in a different, perhaps very distant, country sufficiently to tolerate increases in their energy bills to support such faraway developments. However, whatever big electricity companies may say in terms of how much they may support large renewable installations in the desert, they will require consumers to support the schemes, since the electricity will be more expensive than supplies from conventional sources. Hence, a collapse of 'identity' EM into mainstream EM where the industry organises solutions out of sight and out of mind of the consumer seems unlikely in this case. Scheer may have more of a finger on the pulse of public opinion than may first appear. The explanation for this lies in understanding the differences in the degree of identification that energy consumers will have with different energy projects.

Interestingly, Czisch (2006) points out that, if a supergrid were to be built including renewable energy supplied from North Africa, it would be much cheaper to supply the electricity to Europe from wind farms sited in North Africa rather than solar plant in North Africa – or, he argues, even wind farms sited in the North Sea. Again, however, what happens is governed by the issue of who will pay for the renewable energy plant, and wind farms in the North Sea are likely to be the choice if it is the citizens of Northern Europe who are paying for them. British people will identify with wind farms off their coasts, but not so much, in the foreseeable future at least, with wind farms built in North Africa.

However, it is relevant here to point out that the attitudes of radical environmental groups have also altered as the shift in ecological modernisation processes has occurred. The renewable energy industry retains a different identity from that of the mainstream energy industry, even though, as will be discussed in later chapters, many of the big renewable development companies may be owned by major energy companies themselves. Nevertheless, wind power and other renewable energy technologies have become much more mainstream in recent years, in the sense that their relations with, and ownership by, the mainstream energy industry have increased. Like other engineering industries, it is a multinational industry wherein companies such as Vestas have market shares all around the world and also site production plant in many countries. Renewable energy is less of a social movement today (although, as is discussed by Toke (2011), this still manifests itself in some ways) and much more of a mainstream industry. On the other hand, radical energy-environmentalism has not abandoned all its notions of ecological and political decentralisation. Instead, the emphasis has changed.

The apparent change in emphasis among many green NGO supporters of renewable energy from locally based renewables to (even) intercontinental trade in renewables is not as paradoxical as it may seem at first glance. For a start, emphases among NGOs differ, although even radical organisations like Greenpeace recognise a need for renewable energy to harness the advantages of large-scale infrastructures when this will promote eco-technological change. This does not necessarily mean the abandonment of the concept of decentralisation. As was seen in the section on decentralisation, even this idea is surprisingly fungible and can absorb the notion of even wind farms that are hundreds of MWs in capacity. However, a further point to be made is that in some ways less has changed than may be thought. Solar pv is still seen as a technology that can be sited on the rooftops of ordinary persons' houses, as well as in more industrially sized 'solar farms' on the ground. Much public campaigning by the NGOs focuses on enlarging the possibilities for energy users to become energy producers – or to secure their own energy 'autonomy', as Scheer (2007) would put it. The contribution of residential solar installations may be, ultimately, small to total generation compared with the contribution of industrially sized solar farms, but the residential solar sector is very important in a political sense, since it helps to embed popular identification with the technology.

What is clear is that renewable energy technologies continue to challenge the conventional energy industries, with the aim of replacing them in part or as a whole. A vital part of the discourse that supports this change is ecologically inspired. Hence in these senses renewable industries retain a clear eco-technological identity. What has changed is that the mainstream energy industry is prepared to buy into this vision, but only to the extent that incentives programmes and favourable regulatory regimes for renewable energy continue to be available. Even in this case, this will depend on how far this renewable trajectory can be pursued before there is conflict with the perceived interests of the mainstream industries. As will be discussed in Chapter 6, in some countries the pro-renewable vision of the mainstream industry is greater (e.g. in Spain) than in others (e.g. in Australia).

The inception and continuation of incentives programmes for renewable energy depend very much on continued positive public identification with renewable energy technologies. This is a technology-specific popular identification, and Scheer and others may have an important point to make, that people will only fully identify with the technologies that seem to relate to them. This is as opposed to the notion that consumers will give a blank cheque to the mainstream industry to use renewables as part of a conventional industrial response to energy crises and carbon reduction commitments. That would be in keeping with the type of EM analysis advanced by Mol (1995) for the chemical industry. However a different EM is suitable for this study of renewable energy.

It is surprisingly difficult to pin down a commonly agreed meaning of decentralisation, even on a basic level. However, the concept of decentralisation may have some purchase in helping us understand the notion of eco-technical identity in the case of renewable energy, since the propensity of the public to allow their energy bills to increase to incentivise renewable energy may be influenced by the degree that people identify with the source in a geographical sense. If they are more local, or regional, or at least nationally based, then there may be some toleration of price increases to pay for the renewable energy plant.

Of course, this leads us on to consider how local opposition to renewables, including many wind farm proposals, manifests itself. The earlier discussion illustrates how some environmental analysts have struggled to understand how it is that apparently environmental technologies can be so controversial on environmental grounds. Unfortunately, many analysts seem to operate from what seems to be a Hajer-esque misapprehension that, provided there is enough deliberation, good citizens will walk down the path towards sustainability. There are two theoretical issues here. The first is the institutions of deliberation, which may be designed in one way or another to favour different interests. The second issue, which may be related to the first, is that the possibilities for a successful deliberative outcome depend on the extent to which there is a basis of agreement between the interests involved, including a sufficiently shared appreciation of what constitutes sustainability. As discussed earlier, the attempt to apply Habermas-inspired deliberative approaches to wind farm planning may well be quite a good illustration of the 'difference' critique of deliberative approaches put forward by writers such as Mouffe (1995). Rather than try to explain how interests should be reconciled through deliberation and argument, we could, in this case, be better served by trying to understand how conflicting identities generate planning and developmental controversies. This way we can perhaps think more about how to construct incentives to satisfy the existing interests with their own identities, or at least encourage some interests to support renewable developments more than they otherwise might.

4
California: The Growth of a Renewables Industry

The USA as a topic for discussion for renewable energy is important for a number of reasons. However, in empirical terms much of its importance lies in the fact that by September 2009 it boasted around 30 GW^3 of wind power, not to mention various quantities of other types of renewable energy. US wind capacity had overtaken the previous wind capacity leader, Germany, by a clear margin. As with much of the focus of this book, most attention will be on wind power development, although, as will become clearer later, other renewable fuels will be discussed, including issues surrounding different types of solar power.

The coverage of the USA is broken down into two chapters, one mainly dealing with California in the period leading up to what has been called the 'wind rush' in the 1970s. The second chapter (after this one) deals with the more contemporary renewable energy programmes in four leading US states in terms of installed new (since the 1970s) renewable energy capacity – California, Texas, Iowa and Minnesota.

The emergence of renewable energy in California as a serious energy option is important for various empirical and theoretical reasons. Empirically, the arguments about technology and funding in California in the 1970s and 1980s have significantly influenced later practices and debates. As will be discussed later, Californian deployment of the 1980s is very significant in that it is the first time that wind power achieved the status of being an industry. Theoretically the case of California is important for two reasons: first, because this case throws into doubt the assumption made by the classical 'mainstream' view of ecological modernisation as promoted by Mol (1995), Huber (1991) and others that technology choices are mostly the preserve of mainstream conventional industries, which choose how to adapt their activities in response to pressures for higher-quality environmental standards. The framework proposed in this work for the energy arena is that it is more useful to analyse how a new renewable energy industry has been allied (and in the early stages been merged) with an idealistic, often liberal, environmental movement often in conflict with an existing fossil fuel and nuclear-based electricity industry. In Chapter 2 five characteristics of 'identity EM' were outlined as indicative of such a process. The second theoretical point of interest is that this case study can also be used to question Hajer's (1995) claim that ecological modernisation can be studied without analysing the notion of interests as being distinct from discourse. As was discussed in Chapter 3, there is considerable fuzziness and differing interpretations surrounding the notion of 'decentralisation'. This issue will be explored further and, by contrast, there will be analysis of both discourses and of interests in order to understand outcomes.

This leads us on to the main research questions. First, how can we account for the emergence of the renewable energy industry in California in this period? Second, what was the significance of this for renewable energy in the USA and beyond? Third, how does this case illustrate the theoretical points I have just mentioned concerning ecological modernisation theory and the use of discourse theory and interest group analysis?

Probably the best-known image of renewable energy in California is the large wind farms in Altamont Pass, Tehachapi and San Gorgonio Pass. Indeed, wind power is the biggest focus of the first part of this section on California (solar power is discussed at greater length later), but it is important to bear in mind that other renewables were developed in the 1980s as well as wind power. Indeed, geothermal energy generates around 40 per cent of the roughly 11 per cent of electricity generated by renewable sources in California (CEC 2009), with wind and biomass each generating just over 2 per cent of California's electricity followed by smaller proportions from small hydro and solar power. However, most of this generation was (that is, at least initially) organised by companies that are independent of the major electricity utilities in California, and this is a consequence of the way that renewable energy emerged as a generation option in the 1980s – and that in itself as a consequence of the Californian politics of the 1970s. A key task for this chapter is to discuss the contribution made by California's liberal politics of the 1970s and 1980s to the emergence of wind power as an industry.

Chapters 2 and 3 outlined how ecological modernisation in the case of renewable energy has involved active intervention in the policy, political and, in the case of Denmark, even in the technological sphere, by activists motivated at least in part by the ideals of a social movement idealism rather than solely by motivations of industrial profit maximisation. What became known as the Californian wind rush of the 1980s certainly had a reputation for profit maximisation. The policy instrument that provided much of the underpinning of the programme, tax credits for investments, 'were subject to abuse, and fly-by-night operators were permitted to compile investments, in some cases fraudulent or even criminal' (Starrs 1988, 141). Nevertheless, this wind rush, whatever its faults, erected sufficient windmills, by 1987, to produce most of both the USA's and the world's wind power supply to the grid. It also constituted at that time most of the installed wind power in the USA (personal communication from Paul Gipe, 27 September 2009). The investment credits may have sometimes been abused, but, as will be discussed, the institutions and policies that underpinned the wind rush were put in place mainly at the behest and on the initiative of a social movement of anti-nuclear activists promoting a 'soft energy paths' alternative. The Californian wind rush was, if not essential, at least crucial to the development of wind power as an industry. As is covered in Chapter 3, the initial market for the nascent industry was farmers and then wind power cooperatives until the Californian market emerged from 1982 onwards, encouraged by the administration of the Governor of California, Jerry Brown. As Preben Maegaard, the Chairman of the Renewable Energy Association at the time, comments:

> When you suddenly have a market that is 100 times bigger than the home market, then you have something that's important because that resulted in the industrialisation. Because in 1985 maybe 3,000 units were sold and it grew over the following years until 1987 and by that time the policy of Ronald Reagan had its effect and then the market collapsed completely and all the 20 companies that had been involved in this export went bankrupt except one of them, that was Bonus, they did not. Vestas also went bankrupt. So you see, and then we had a production capacity that was many times more than the market was because then there is a difficult period from say 1987/8 to 1991 when the German market opened up. (Interview with Preben Maegaard on 18 April 2009)

The importance of 'non-material' politics

In order to understand the background to the wind rush it is necessary to bear in mind the existence of a movement in California opposing construction of large utility power plants, focusing most of all on nuclear power plant. It was a movement that began as early as 1958 with the controversy about a nuclear plant proposed for Bodega Bay, and continued through the 1970s (Wellock 1998). A long series of sites for nuclear power were proposed and then abandoned in the face of the opposition. Wellock (1998, 6) states: 'unlike cancelled (nuclear power) projects elsewhere, none of these plants failed owing to slumping electricity demand, construction costs or federal regulatory delays.' In 1976 California adopted a 'Nuclear Safeguards Act', which, by imposing a moratorium on new nuclear construction until a solution for nuclear waste disposal was found, effectively ended possibilities for constructing nuclear power stations in California.

Wellock argues that 'New social values as much as any other factor brought the antinuclear movement to life...As the authority of public officials declined in the seventies, the antinuclear movement's message spread out beyond its environmentalist moorings to appeal to populist elements. These groups mounted their own critique of the liberal state and were drawn to the movement's anti-authoritarian, anti-federal rhetoric.' This movement appealed to 'non-materialist' values. 'Such non-materialist values were especially popular among young, well-educated white Americans who valued amenities more than economic growth more than technological progress' (Wellock 1998, 8–9).

The emergence of this movement can be anchored even further back in the traditions of environmentalism in California, which gave birth to the Sierra Club. The Sierra Club was formed in the context of theoretical and practical debates about the nature of environmentalism. The Sierra Club earned much of its early reputation at the beginning of the twentieth century out of the (unsuccessful) campaign, led by John Muir, to stop the 'Hetch Hetchy' dam project being built in the Yosemite National Park in California. Yet, as with other battles lost, it created a movement and shifted dominant discourses towards greater stress on preserving areas identified as aesthetically beautiful or consisting of 'wilderness'. The debate between the aims of conservation as being, as Muir argued, for preservation *per se* or, as the Head of the Forest Service, Gifford Pinchot, argued, for the best practice path towards economic development remains at the heart of environmental debates in more contemporary times (Eckersley 1992, 39–40; Norton 1994, 17–38).

The attachment to conservation formed a strong Californian tradition that has acted as a powerful influence shaping debates and practices about environmental issues, including energy regulation, right up to today, and will no doubt continue into the future. As Bevir and Rhodes

(2003, 34) point out, political outcomes can be explained in the way that traditions 'show how individuals inherited beliefs and practices from their communities', and also in the way that traditions are interpreted and reinvented in the light of new policy dilemmas. Opposition to large energy installations and support for conservation of nature for its own sake are two very strong traditions in California. The emergence of the movement for what can be loosely be termed the 'solar alternative' can be seen as a continuation of traditional Californian environmental approaches, reinterpreted in the light of the dilemmas posed by the energy crisis and the liberalising impact of the anti-nuclear movement.

Of course, even when it comes to debates about land planning issues in California in the 1970s, and elsewhere, debates seldom break down into the distinctions used by Muir and Pinchot. Rather, 'non-material values' as Wellock (1998) puts it, mean something more fuzzy than absolute attachment to a pristine environment. They include concern for conservation of biodiversity, but also for more human concerns that take into account longer-term issues, such as the future impact of energy resource depletion or global warming, and not merely the short-term interests of industry. This can be incorporated into what Barry (1999) calls an 'enlightened' notion of human interests, which seeks liberal environmental objectives to protect the ecosphere for human purposes even though, for example, there may be no specific knowledge about the importance of a particular habitat for human interests.

Of course, this works the other way around as well. Campaigns to protect individual species or to protect specific aesthetic environments can sometimes be interpreted (or at least perceived as interpreted) as being raised in support of existing human economic interests. Certainly environmental organisations have been involved in such arguments, and sometimes dilemmas in the case of the siting of wind power plant. It has certainly involved the Sierra Club in California in such dilemmas. They have often expressed support for renewable energy, as well as their preferred option, energy conservation, but have also had a record of sometimes supporting campaigns against specific wind farms, including the classic case of the Tejon Pass proposal in 1987 (Gipe 1995, 446–50; Righter 1996, 236–8). This discussion of 'values' is central to the discussion of how ecological modernisation has worked in practice in the development of renewable energy. As was discussed in Chapter 2, Mol (1995), representing the 'mainstream' approach to EM, examines the chemical industry's responses to ecological pressures, but does not study the emergence of alternative industries (e.g. organic agriculture) that are, initially at least, independent of the establishment corporations that manufacture pesticides and other products. The mainstream version of EM sees value change as part (albeit a crucial part) of the context that induces existing industry to adapt to new sets of values. However, in order to understand the operation of EM in this case we need to understand value change acting as a handmaiden to the birth of a new industry which gave opportunities for entrepreneurs that were independent of the existing dominant electricity industry. A 'non-materialist' movement did a lot more than pressure an existing industry to come up with new techniques to respond to environmentalist pressures; it discussed and promoted independent technological paths, including that of wind power.

This allowed the development of the wind industry. In short, without the forces arguing for a 'value change' there would have not been sufficient support for the institutional, regulatory and financial incentives that were necessary to bring the new industry into existence in the first place. Hence more attention needs to be focused on the movements arguing for value change, and the ways in which a new industry emerged, rather than focusing on how the existing electricity industry reacted to the environmental and resource pressures in California of the 1970s and 1980s.

We cannot study the emergence of the renewable energy industry in California by spending most time studying how leading utilities such as Pacific Gas and Electric or Southern California Edison altered their policies to give contracts to renewable energy developers. The utilities had to be dragged under great pressure to give contracts, and projects only emerged because of state-level and federal-level incentives and institutions that had emerged during the 1970s. Mainstream EM can be better applied to the process of how the utilities, under pressure from environmental groups such as the Sierra Club and the Environmental Defense Fund (EDF), adopted energy conservation as an important strand of their programmes. However, the development of wind power in the 1970s and 1980s in California was driven by a combination of incentives supported by liberal anti-nuclear activists and the activities of a number of independent wind power developers who took advantage of these incentives. Some time will therefore be spent on discussing these issues.

Towards a solar society?

The motivating force that put the pro-renewables policies into place in California crystallised into political forms in the early 1970s. An

anti-nuclear movement had, as Wellock (1998) implies, become associated with other anti-establishment political forces of the time,
including the student-based anti-(Vietnam) war movement, and such
opinions swelled the ranks of what became a liberalised Democratic
Party in California. This movement supported alternative approaches
to the menu of large conventional power plant (including nuclear
power) offered by the electricity utilities. Of course, this movement had
different emphases. The more politicised, left-wing elements, such as
Tom Hayden and the Campaign for Economic Democracy, tended to
be more consistently supportive of renewable energy than the Sierra
Club. The Sierra Club, who were influenced by aesthetic as well as other
environmentalist motives, tended to favour conservation rather than
renewables, although for the period of the high oil prices after 1973
they became more sympathetic to the renewables agenda. As discussed
in Chapter 3, during this period it was less plausible to argue for a non-
nuclear strategy that involved only conservation of energy rather than
alternative supply options.

Another aspect of the political coalition for renewables is that it
drank heavily from the ideologies found in the writings of Lovins and
Schumacher. These soft energy path, decentralised, 'small is beautiful' notions fitted in well with the anti-authoritarian and anti-utility
feelings that were rife in the anti-nuclear and anti-war movement. In
the 1970s and early 1980s the 'soft' energy path was generally associated with a move towards a 'solar society'. 'Towards a Solar Society'
was the call issued by Hayden himself (1980, 83–124), and the term
'solar' seemed to be used to cover most renewable energy, conservation
and cogeneration for local use. In pursuing this strategy Hayden's scepticism, sometimes outright contempt, for the energy utilities and the
military industrial complex was clear. He said:

> The question is whether the corporate system can undertake and
> survive a rapid transition from depletable energy sources to renewa-
> ble ones: the real test is whether the present corporate arrangements
> can 'get the job done'. The goal has to be the energy transition,
> not the preservation of current market structures ... largely aiding
> and abetting the giants with contracts and alluring policies has
> been the Department of Energy ... (whose top 20 officials had) a
> combined employment experience of 209 years in the Department
> of Defense, Atomic Energy Commission, Central Intelligence
> Agency, National Aeronautics and Space Administration, and mul-
> tinational oil firms. Trusting men like these with solar energy is

akin to putting Dracula in charge of the community blood bank. (Hayden 1980, 121)

Hayden's denunciations of the energy establishment for their failure and inability to advance the cause of developing soft energy solutions may be over-colourful, but, leaving aside some of the emotive invective, there is a theoretical lesson to be drawn here that follows from the theme about EM. Does the Californian case illustrate the mainstream EM case that mainstream industry is the main handmaiden of ecological technological change? The evidence, as discussed in this chapter, suggests that, at the start of the process at least, the answer is more likely to be 'no' rather than 'yes'. Yet the dream held fondly by Hayden and other 'small is beautiful' advocates of a great upwelling of community adoption of soft energy paths seemed not to emerge (in California) either. Hayden declares that:

> The transition to a conserving, solar society can and will occur through a renewed spirit of community self-sufficiency...There are signs of this change everywhere...The society based on renewable resources rises into vision through the efforts of its pioneers. The basic change comes on a personal and community level, and can neither be controlled nor taken away from above. (Hayden 1980, 122–3)

Hayden cites various examples of local poor and ethnic minority groups developing, in a technologically inventive way, their own solar projects. There were a few examples of these activities, but it has never amounted to a mass movement in California. That such things are possible is signalled by the example of Denmark, which, as is discussed in Chapter 3, not only virtually invented modern wind power, as well as helping to invent farm-biogas generation, but also developed an extensive community-based cogeneration system that today supplies more than half of the electricity and heat in that country. The Californian liberals may have talked more ideology, but the Danes did the practical work to develop alternative technology. As is discussed in Chapter 3, the ideologues could not satisfy the pragmatic Danes because they did not talk about weldings, pumps and types of fibreglass blades, which interested the farmers and blacksmiths who made the machines (Interview with Preben Maegaard on 18 April 2009). Again we can explain such differences by reference to different cultural traditions. The Danes had a tradition based on rural technological 'DIY' self-sufficiency and cooperative ways of organising communities. By contrast, the Californians had a strong ideological tradition of nature conservation and a more recent liberalised anti-authoritarian, anti-nuclear movement.

There were many activists attracted to promote ecological 'appropriate technology' (AT) alternatives. For example, during the 1970s Lisa Daniels, who is a leading wind power organiser in Minnesota today, worked at the Farallones Institute, based in Berkeley, which was concerned with developing self-sufficient urban building designs and other sustainable technologies (interview with Lisa Daniels 11 May 2009). The Institute was founded by Sim Van der Ryn, who served as California State Architect under Jerry Brown and was involved in the management of the Office for Appropriate Technology. Although such things did not resolve into much practical, grass roots-based innovation, such initiatives were important in spreading intellectual ideas to future industrialists and policymakers, and as such they do form an important part of the ecological modernisation process. However, a major contribution to the EM process resides in the fact that the AT or 'small is beautiful' movement provided essential support and inspiration for the incentives established to allow the deployment of renewable energy on an industrial scale in the early 1980s.

As the 1970s wore on there was growing resentment among representatives of such political trends that the notion of 'appropriate technology' was being stretched to include proposals to build corporate-inspired large-scale renewable energy ventures. Sim Van der Ryn criticised plans for large-scale solar electric power stations (Greene 1978). In fact, some solar thermal electric power plant was built by Luz Corporation in the 1980s under the renewables incentives progamme, and a total of 384 MW was installed in the Mojave Desert (Brower 1993, 52–3). What is interesting is that such types of plant, up until now, have not been developed very widely. Solar thermal electric has been regarded as being too expensive to be economic with the incentives available to wind power, yet it is cheaper than photovoltaics, which have been deployed more widely, despite their cost. One explanation for this difference is related to the politics of grass roots support. Solar pv can be, and in several countries, led by Germany, has been, installed by individuals on their rooftops. As discussed in Chapter 6, this institution of owner-generation creates its own populist interest group, which, in collaboration with a strong coalition of interests, has led to high levels of subsidy (usually in the form of feed-in tariffs) paid for electricity generated from solar pv. By the end of 2008 over 5 GW of solar pv capacity was installed in Germany alone (Reuters 2009). Large solar thermal electric technology does not have this type of political support. As will be discussed in the next chapter, the incentive structure under the Californian RPS has changed, and a surge in orders for solar thermal electric plant in California is now happening. However, as will also be discussed, things may not be as straightforward as the numbers of MW in the contracts issued suggest.

As has been said (and something which needs to be emphasised), the fact that the Haydenites and the 'small is beautiful' advocates did not achieve their stated aims does not minimise their impact in other ways in developing institutions and incentives that allowed an 'alternative' energy industry (to the established energy utilities) to be launched using the entrepreneurial skills for which Californians are well known. Let us briefly examine the institutions and incentives established as a result of the pressure from the followers of Hayden (and others).

Jerry Brown and the push for renewables

Charles Warren chaired an Energy Sub-Committee of State Assembly which proved crucial in shaping California's energy policies in the 1970s. He said:

> Under the rules of the Public Utilities Commission, they (the electricity utilities) were assured a guaranteed rate of return on that investment so there was a strong motivation in the utility industry in California to increase demand. Therefore we had already KW programmes urging consumers to use electrical appliances and little attention being given to efficiency techniques, because they would make more money. In late 1960s, it became clear that there were pressures building, making it increasingly difficult to increase generating capacity. First, most of the hydro electric resources had already been developed, coal could no longer be used in California because of air quality concerns, the supply and cost of oil was troublesome and occasionally unpredictable. So the utilities began to increasingly look to the nuclear option. (Interview with Charles Warren, 27 October 2009)

However, opposition to the nuclear option increased steadily throughout the 1960s and 1970s. In 1976 the State Assembly passed legislation that put an effective moratorium on construction of new nuclear power stations until the problem of storing nuclear waste could be solved. This moratorium has never been reversed.

The effect of the double coincidence of the strong, liberalised, antinuclear movement and the inception of the oil crisis in October 1973 can be seen in the debate in the Californian Legislature concerning the establishment of what became the California Energy Commission (CEC). Before the oil crisis, a proposal to establish what became known as the CEC was put forward from the floor of the California State Assembly by Charles Warren. Its aim, according to its environmentalist proponents, was to democratise the process of dealing with proposals for new energy plant and incorporate environmentalist objectives into the process (Wellock 1998, 117). Proposals for the CEC were approved by the State Assembly. Then, just a few days before the outbreak of the 'Yom Kippur' war, which preceded the Arab oil embargo (and the first oil price spike), Governor Reagan vetoed the Bill setting up the CEC. However, in the aftermath of the oil crisis, heightened concern about the need for an energy alternative was associated with Reagan's acceptance of the second effort to pass the Bill establishing the CEC (Wellock 1998, 142–5).

Jerry Brown succeeded Ronald Reagan as Governor in 1975, and served until 1983. Brown already had a record of cooperating with liberal Democrats in the early 1970s, and he was elected during a wave of reaction against the Watergate affair. Although he came into office without firm convictions on energy, he interpreted his interests as being served by promoting what were seen as the 'soft energy path' goals that were supported by the liberals. In 1976 Tom Hayden ran a close second to the Democrat incumbent, John Tunney, for a seat in the Federal Senate, something which was seen as contributing to Tunney's defeat later on. Brown wanted to avoid Tunney's fate and preferred having the support, rather than the opposition, of what was then a powerful liberal constituency among Democrats in California (Bunzel 1983, 14–16). From 1976 onwards (until he left office in 1983) Brown placed a great emphasis on promoting the alternative energy agenda. He appointed people who were sympathetic to the aims of the alternative energy agenda so as to constitute a majority of the Public Utilities Commission and the CEC, and he established the Office for Alternative Technology (OAT).

Brown studied alternative energy issues with great vigour, and placed considerable emphasis on the work of Wilson Clark, who tended to emphasise the benefits of large-scale renewable energy as well as the small-scale, off-grid tendencies favoured by Lovins (1977) and the 'small is beautiful' thinkers. Clark was appointed as Brown's energy advisor. Clark was a keen supporter of wind power, writing in 1974 that: 'Not only is the technical development of electric wind generators encouraging, it is revolutionary...The minimal drawback of higher capital investment for wind plants would be more than repaid by the benefits of the inexhaustible energy, safety, and environmental compatibility' (Clark 1974, 563). Clark's writings have a less polemical ring about them than Lovins's. When Clark talked about decentralisation, he discussed the technicalities and possibilities of technologies such as cogeneration (called combined heat and power in Europe) and also combined cycle gas plants, which were then a new concept. In that sense he was prescient in a conventional power sense, as well as in promoting energy conservation, renewables and other devices such as heat pumps. Despite the publicity given to the wind farms constructed in the mountain passes in California in the 1980s, the biggest gainers in terms of capacity were the oil and gas-fired cogeneration plants, which benefitted from the policies that flowed from the coincidence of Federal policy changes and the pressures emanating from Governor Brown's office.

Clark was hired along with a group of 'small is beautiful' advocates such as Gigi Coe, Bob Judd, Kirk Markwald and Ty Cashman to join a new organisation established by Brown in May 1976 called 'The Office for Appropriate Technology' (Interview with Jan Hamrin 15 April 2009). The aims of the organisation ring with Schumacher-style language:

> The recognition that we live in a world of limited resources requires development of a conserving technology. As government tries to adapt to the new realities of diminishing resources and changing values, we must find ways to carry out our responsibilities in ways that are less wasteful, less costly and bureaucratic, less costly to people and our environment. We need to encourage tools, techniques, and processes – in our economy as well as our communities and our institutions – that are simple, direct, small scale, and inexpensive: a balanced technology that is appropriate to maintaining the health of California's people, economy and environment. (Pursell 2001, 306)

The OAT, in keeping with the views of Tom Hayden and the spirit of Schumacher (1973), initially opposed notions that the utilities could have a role in developing alternative energy sources (Roe 1984, 118–20; van Est 1999, 35–6). This view coincided with the failure of the California legislature to endorse the CEC's plans for funding demonstration large-scale wind turbines. 'Small is beautiful' adherents wanted development of small wind turbines, not large wind turbines manufactured by multinational companies (van Est 1999, 39–40). Such turbines were being built by companies such as General Electric and Boeing, who designed the demonstration MOD 1 and MOD 2 machines for NASA and the US Government. Yet, they did not work (Gipe 1995, 103–7). As is discussed in Chapter 3, these machines (and other large devices funded by government programmes) failed in all cases of government R & D programmes precisely because the large utilities insisted on large machines that were, at that time, impractical owing to a lack of learning about how to increase wind turbine size above the scales that were then known to work. The instincts of the 'small is beautiful' activists (in being sceptical about the efforts of the multinationals to design large wind turbines that worked) were right, even if the reasoning was put in ideological rather than engineering terms. Despite such foresight, the OAT was disbanded in 1983 after the election of a conservative successor to Jerry Brown as Governor of California.

On the other hand, it seems that a Mol–Huber 'mainstream' interpretation of ecological modernisation might involve a strategy of encouraging the multinationals to develop wind turbine technology. Indeed, this is what is happening in the twenty-first century, as is discussed in later chapters, although even here the story also involves a great deal of activity from independent and even (mainly in parts of Europe) a lot of community-based developers. However, in the 1970s and early 1980s it was a rather different story. It was Danish engineering/farmer enthusiasts and small companies that developed the technology, as described in Chapter 3. It was the Danes who made machines that worked on the basis of an idealistically based competition based on open information. The development of industrial standards in the 1980s meant that the machines could be incrementally increased in size as the engineering issues were solved, again on an incremental, experience-based and market-tested basis.

This illustrates a theoretical point about EM, in that the liberal ideologists as well as the Danish enthusiasts had an irreplaceable role in the development of a new industry, that of wind technology. This was done in the face of opposition from most of the existing electricity industry. In the case of the Californian 'small is beautiful' liberals, this role was in terms of their pressure leading to the provision of an effective incentive structure for decentralised entrepreneurial competition to sort out the better from the worse wind devices. The wind farms established in the 1980s in California were exercises in Schumpeterian 'creative destruction' in that what was partly a demonstration programme pointed the way forward towards industrial development by Danish manufacturers such as Vestas, while other designs fell down, often literally.

Developing institutions

In fact, the ideology of the OAT was tempered in practice by the need to offer suggestions that would fit into the existing matrix of institutions. They advocated programmes to support renewable energy and energy conservation at all levels, Federal and State, and by both utilities and independents (Office of Appropriate Technology 1980). A major impact of the renewable energy proponents who ran the CEC and the OAT was the practical resources and their support for a system of incentives for renewable energy. The CEC prepared wind resource reports, which enabled independent developers to site projects in windy areas and raise capital. This in itself was a resource that was not available in other states. Even though other states might, in physical terms, be windier, in sociological terms only the wind speeds on many Californian sites counted as 'knowledge'. Wind power went to California, and not elsewhere (Righter 1996, 205).

Then there were the tax credits, which enabled investors to write off capital investments against their tax liabilities. This began as a 5 per cent tax credit for solar power devices, but was ratcheted upwards and broadened to include wind power (Interview with Jan Hamrin, 15 April 2009). By 1978 this had transformed into a state-based 25 per cent tax credit. To this could be added a 25 per cent tax credit for renewables that accumulated under the combined headings of: 1) a regular (10 per cent) investment tax credit for business investments and 2) (15 per cent) energy tax credits under the Crude Oil Windfall Profits Tax Act passed in 1980. On top of all this, renewable energy generators could also qualify for 'accelerated depreciation benefits', although this provision extended to a range of business investments (Starrs 1988, 116–19). The combination of these allowances meant that the investors could claim relief not only on most of what they had invested but also on the money they had borrowed to finance renewable energy projects. According to Alvin Duskin, designer of a Federal tax credit amendment pushed through Congress, as a result of the various incentives 'In essence the investor [got] all of his profits for nothing' (Asmus 2001, 82). Any income for generation could, according to this logic, be seen as pure profit. Of course, given the uncertainty about the technology at that stage, it needed investment costs to be almost completely subsidised to encourage investors to put up the money. This was a matter of creating confidence in the possibility of returns where otherwise there would be none. Of course, as we shall see later, the existence of contracts clearly stipulating income levels proportionate to the number of energy units produced was also necessary to instil financial confidence in the renewable projects.

Many criticisms have been made of this system of incentives, especially the arguments that it allowed some actors to make excess profits out of selling on machines. In addition, the investment tax credit approach put the emphasis on installation of the equipment rather than on reliability and production of electricity. However, it certainly worked as an incentive device when allied to another set of incentives that could guarantee returns for each unit of electricity that was actually produced. These production incentives flowed in the wake of the federal Public Utility Regulatory Policies Act (PURPA), which was contained in the National Energy Policy Act in 1978. This obliged the incumbent monopoly retail electricity utilities to offer contracts to independent generators offering payments equal to the 'avoided cost' of the generation. Avoided cost is what it would cost to obtain the power from other sources. This was subject to a wide basis for interpretation at a state level, and the usefulness of this legislation to cogenerators and renewable energy developers largely depended on the extent to which state-based regulators were prepared to enforce it and interpret it in favour of the independent generators.

We can see that an array of institutions was being established to promote renewables. These included different incentives and organisations, such as the CEC and the OAT, which offered professional support and information about wind resources. However, this was not quite enough to ensure that independent generators were able to put projects into operation. Another important institution that was so far lacking was the provision of contracts to the generators that would allow them to sell their power to the electricity suppliers.

By the end of the 1970s the California Public Utilities Commission (PUC) had a majority of Commissioners (appointed by Brown) who were favourable to Brown's strategy. PUCs have the task of regulating what are legalised electricity monopolies. PUCs vary between states in their degree of intervention, but under the Brown Governorship the California PUC became very interventionist in favour of energy conservation and alternative energy supplies. The 'Jerry Brown' Commissioners, pushed on by public interest groups such as EDF, created sufficient motivation to induce the electricity utilities to support an alternative programme to that of conventional power stations, which included renewable energy sources as well as energy conservation and cogeneration (Roe 1984, 187–9).

Initially only well-organised companies that could hire good teams of lawyers, such as Zond and US Wind Power, managed to climb on top of the paper mountains required to negotiate long-term contracts (power purchase agreements, PPAs) with the utilities (Asmus 2001, 104–10). It was only after Jan Hamrin organised the formation of the Independent Electricity Producers that standard contracts for renewable generators, which came to be called 'standard offer four' or 'SO#4' contracts, were negotiated. Hamrin had earlier promoted a speaking tour by Amory Lovins and had worked as a policy officer for the CEC, being responsible for the formulation of investment tax credit schemes for renewable energy. The negotiations for the SO#4 contracts were backed by pressure from the California PUC and took place in 1982. The PUC gave an order to the utilities to issue such contracts and threatened them with fines if they did not comply. These contracts offered fixed standardised payments for power produced for 10 years, although in the case of cogenerators there was also indexation with the price of fuel. According to Jan Hamrin (interview 15 April 2009) the basic contract design 'looked a lot like a feed in tariff, it had a fixed price and it had standard contract terms and conditions'. That having been said, the contracts look odd by today's standards of feed-in tariffs in that the prices paid actually went up over the 10-year contract period, rather than down, as in the case of the classic model of feed-in tariffs in Germany (where contracts run for 20 years).

These contracts allowed a great expansion of cogeneration and renewable energy. Among this there was, by 1987, 1500 MW of wind power installed. This was joined by large quantities of biomass, geothermal, small hydro and also some solar thermal electric plant. As was implied earlier, the generally superior performance of machines from Denmark (where a small rural market had already been developed) cleared the way for the emergence of a Danish wind power industry that still enjoys world dominance today. Some of the industry, such as the wind turbine manufacturer BONUS (now owned by Siemens), has been taken over by other companies since the 1980s, although Vestas remains the world's largest manufacturer and is still a wholly Danish company.

Initially the wind development was owned by independent companies. Zond, for instance, led by Jim Dehlsen, developed much of the wind capacity at Tehachapi, although later much of Zond's holdings were sold off to Florida Power and Light (FPL) (USDOE 2003). For many years California boasted the lion's share of the new renewable energy installations. However, little was deployed in the 1990–2007 period. It is only very recently that the impact of the RPS legislation is now being seriously felt in terms of increasing renewable capacity. More will be said on this subject in the next chapter.

Popular discourses and interests

As mentioned earlier, discourses and interests often need to be analysed separately. As argued in Chapter 1, we should not assume, as implied sometimes by analysts such as Hajer (1995), that interests in environmental controversies can be assumed to be little more than subsets of identifiable environmental 'storylines'.

A study of storylines is often essential, but the primacy accorded to them over analysing how actors pursue interests may sometimes be misplaced. One factor is the sheer ambiguity of some storylines. The earlier quote from the principles of the Office of Appropriate Technology gives a notion of an important storyline during the Jerry Brown era, although, as is seen from the previous discussions, including those in Chapter 3 about the notion of decentralisation, the signifier 'decentralisation' in fact has somewhat diverse meanings attached to it. There is an immediate problem of what it is that constitutes the storyline in this case. Was it against any relationship with utilities? What was the relative emphasis given to conservation, different renewables and cogeneration? Just how small was 'small'? The writers most associated with the alternative energy movement themselves seemed to have different emphases. Holdren and Herrera (1971), published by the Sierra Club, gave most emphasis to conservation and relatively little to supply-side options such as wind power; Clark (1974) gave much more scope for large-scale renewables in general; while Lovins (1977) gave great emphasis to an anti-nuclear, anti-large-scale energy narrative.

The energy alternative narrative was fuzzy in terms of whether its priorities were mostly energy conservation, small-scale renewables, large-scale renewables, some or other mixture, or even schemes like the solar generating satellites that were then (and now are again) items of renewable speculation (some would say fantasy). This allowed a number of different organisations to support the incentives for alternative energy that were promoted through this narrative. Upon analysis, however, it is difficult to argue that we can solely understand what happened by assuming that these organisations were, as Hajer implies, contingent to a storyline rather than also by posting them as having distinct interests. Moreover, if we are to understand the practices of actors that supported the energy alternative storyline, it is difficult to do so without being specific about what these groups' interests were in particular contexts. These interests are at their most specific when they involve financial issues. As Rhodes (1997, 20) implies, academic analysts can, using an interpretive approach, still apply agreed analytical concepts to different case studies and different narratives. I use this style of reasoning to support use of a conception of interests to enable us to analyse how actors perceive their own interests and act to advance them. Given that interests of various participants in the Californian energy debate are perceived to be different, and that the energy alternative storyline is itself 'fuzzy', it seems analytically fruitful to analyse practices and outcomes in terms of interests, regardless about arguments about their existence as 'real' objects.

As implied in the following explanation, one can analyse two distinct 'energy generator' groups of actors wishing to benefit from the energy alternatives discourse:

> The federal government passed a renewable energy tax credit bill during the same session of Congress as the one that passed PURPA. As I understand it, from the people who were involved in drafting these two pieces of legislation, no one realized the synergies between the two. On the one hand they saw utilities unwilling to build co-generation or renewable power and felt they had to do something to influence that behaviour. That was the reason for PURPA. On the other hand they saw a potential flaw for individuals wanting to build renewables because they are so capital intensive. Mom and Pop own a farm and want to put up a little windmill or a little hydro turbine in the stream on their property. A tax credit could provide that extra financial incentive and support. The folks that wrote the PURPA bill and the ones that wrote the tax credit legislation have given me the impression they never saw the two bills going together. But the financial community connected the dots and folks who had been financing condo conversions and real estate all of a sudden thought 'Wow, we could finance renewable energy and the utilities would have to buy it.' So that connection was made and that caused a tremendous interest by the financial community for investing in renewables. Jan Hamrin (interview 15 April 2009)

The gate was opened by the energy alternative storyline, and the incentive structures that followed in its wake, for organised commercial interest groups to form. Initially, in 1974, the American Wind Energy Association (AWEA) was formed by 'small is beautiful' adherents, but later, especially upon the passage of the PURPA legislation, it became the representative of the commercial developers. The Independent Electricity Producers Association was also formed in the wake of PURPA. These represented interests that were distinct from the latent army of residential and small farmer generators imagined by the 'small is beautiful' advocates, but which failed to materialise in the 1970s and 1980s in California. The commercial energy developers had their own distinct interests for long-term contracts, and the lobbying and outcomes must be analysed with this specification of practices in mind. Such practices did not automatically flow from the 'small is beautiful' version of the energy alternative storyline, and must be analysed separately as an interest.

Important actors which, for a time, bought into the energy alternatives storyline were many energy consumers represented by groups established to promote energy consumer interests, for example Towards Utility Rate Normalization (TURN). They were open to arguments that energy alternatives could reduce what were seen in the 1970s as escalating conventional energy generation prices (Interview with Jan Hamrin, 15 April 2009). However, the support of energy consumers tailed off as energy prices fell after 1985. This made it easier for a counter-storyline, pushed by utility executives, against further subsidy programmes for what they saw as excessively expensive and unreliable alternative energy supplies. In early 1987, according to Melloan (1987), Dick Clarke of PG&E said:

> PURPA requires us to buy power produced by so-called independent power producers. We've been required to execute contracts to buy power even if we don't need it. It's a very expensive resource; we can't dispatch it and we can't count on it. So we have all these contracts for something like 9,000 megawatts. Our whole system has only a 15,000-megawatt capacity, so these contracts represent 60% of our energy capacity... Now so far, only 1,600 megawatts have been constructed; the rest are just pieces of paper. But even if only 45% of the other 7,000 megawatts come on line, it will be an additional cost to our customers of $850 million a year.

Again, as with the interests of the energy consumers, there is a certain independence of interests from what had been a dominant storyline of support for energy alternatives in the late 1970s and early 1980s. Utilities may have found it necessary to bend to legislative and regulatory pressures to allow independent generators to have contracts for a period, but their own perceived interests of earning returns on investments in power stations for which they seek permits and for which they have knowledge and expertise. As the 1980s wore on, and after oil and other energy prices fell in 1985 and 1986, support for the investment tax credits for renewable energy at both California state and Federal level ebbed, and the incentives were ended as the utilities' own storyline gained wider acceptance. The renewable energy installation programme came to a virtual halt.

The CEC and the PUC later argued for more contracts to be granted to renewable energy, but this was ultimately resisted by the utilities on the grounds that it interfered with the process of electricity restructuring and 'deregulation' that occupied the electricity regulatory agenda in the mid-1990s. As we shall see in the next chapter, a storyline about competition and 'deregulation' became dominant.

Conclusion

The first research question posed at the beginning of this chapter was about explaining the emergence of renewable energy in California. This emergence occurred at a coincidence of a) a long-standing tradition of environmental protests and conservationist action in California; b) a sustained series of campaigns against large power station, especially nuclear power, planning applications; and c) the onset of the oil crises in the 1970s.

Using Bevir and Rhodes's (2003) metaphor, traditions of conservation have been extended to campaigns against nuclear power. This, however, left anti-campaigners with a dilemma, in that alternative energy sources had to be found. At first the emphasis was on energy conservation, but the oil crises sharpened the dilemma concerning how energy services were going to be supplied. This increased the arguments for renewable energy, empowered those arguing for incentives to be given to renewable and independent energy suppliers and led to the formation of industrial interest groups arguing for these options. Other political factors were important, including the liberalisation of anti-nuclear opinion through anti-Vietnam war protests and the emergence of political leaders (in particular Jerry Brown) who were willing and able to harness this political constituency and deliver to them policies promoting an alternative energy strategy.

The significance of the Californian renewable energy programme was far-reaching for many reasons. These reasons included the bringing into existence of a significant quantity of renewable energy generation capacity which could be used as evidence that the renewable energy part of an alternative energy strategy was plausible. The wind turbines in the Californian mountain passes quickly became an iconic representation of the alternative energy strategy. However, a very practical significance of the programme was that it created, for the first time, the possibilities, experience and example of having a wind industry on an industrial scale. Although Danish wind generator manufacturers went bust as the wind market ebbed in the late 1980s, important seeds had been sown, which spurred on development in European countries.

We can analyse the programme as occurring as institutions emerged that could satisfy the interests of the nascent renewables energy industry. Incentives, contracts to supply electricity, and the support agencies were key institutions in this process. In the next chapter the theory of institutions will be expanded. For now, it is important to explain how this case illustrates the point that ecological modernisation should not be reduced to a study of how established industry accommodates to emerging environmental pressures. In this case a new industry, or new set of industries, emerged, which were independent of the existing energy utilities. As we saw in Chapter 3, an anti-nuclear movement in Denmark, operating from the grass roots, virtually invented modern wind power (and also farm biogas) technology. California presented the new technology with a much bigger market, thus making an important contribution to the development of a mainstream wind power industry in later years.

Van Est (1999) recognises that both Danish and Californian experiences can be described as ecological modernisation. What needs to be added here is that the nature of EM in these cases is rather different from that described by mainstream EM analysts such as Mol (1995). In both cases, albeit in different ways, actors outside the conventional energy corporations and interests have been instrumental in making and implementing energy choices that were not initially favoured, and not implemented by the electricity utilities. Again, in contradistinction to the emphasis of Hajer (1995), this popular involvement has been about technological issues rather than issues of deliberative design.

Hence, we can advance an argument that ecological modernisation in the case of renewable energy involves all of the five characteristics of 'identity' EM outlined in Chapter 2. First, there were idealistic objectives and activism associated with institutions such as the Office for Appropriate Technology. Second, a renewable energy industry emerged. This industry was independent of the existing energy utilities and emerged largely in the face of their opposition. Third, an incentive structure emerged to support renewable energy, involving a mix of state and federal-based tax incentives and action by California state-appointed regulators to induce utilities to offer long-term power purchase agreements to independent renewable energy developers. Fourth, there were coalitions between environmental (and consumer) NGOs in support of alternative energy programmes. Fifth, independent renewable trade associations emerged during this period.

The development of renewable energy in California in the 1970s and 1980s may have borrowed much of the technological insight from Denmark, but a social movement was essential to providing the initial administrative infrastructure and financial support to create a wind power industry, which hardly existed beyond a part of the Danish agricultural sector (Toke 2011). Although the importance of idealistic activism may have declined in the later stages of development of the renewable energy industries, as will be discussed in the next chapter in the cases of states such Texas, Iowa and Minnesota, there is still today much evidence for other characteristics of 'identity EM'.

As regards the use of discourse and interest group analysis, another focus of Hajer (1995), we can see from this case how it is important to analyse how actors interpret the usefulness of different storylines to their own perceived interests. Consumer interest in renewable energy rose and fell according to their perceptions of whether it could reduce their costs, and different, opposing interests attempted to mobilise these changing interpretations.

This is in addition to analysing how discourses have changed actors' perceptions of their own self-interests. Renewable energy generators benefitted greatly from storylines about the need to incentivise solar power, but they need specific structures to give them confidence in future returns, regardless of 'storyline'. Moreover, discourses of how incentives should help small, local, renewable energy schemes were mobilised by more industrially oriented, if independent, renewable entrepreneurs to seek support for their own ventures. As we shall see in the next chapter, a precise specification of the interests of different actors, as well as an understanding of dominant discourses and storylines, is essential if we are to understand the energy outcomes.

5
USA: Consolidation of a Renewables Industry?

In the last chapter I discussed the birth of the renewables deployment programme in the USA, and its significance for the wind power industry, which consisted of the developments in California in the 1970s and 1980s. Indeed, as Paul Gipe has emphasised (personal communication, 27 September 2009): 'The myth of California being a renewable mecca is 25 years old and that the state has been living on its laurels ever since.' In fact, as we shall see as I continue the California story up to the present day, California is losing the renewable development lead, and there are some question marks over the plausibility of some aspects of the current strategy. It follows from this that other (currently) leading states where renewable development has been gaining pace deserve closer attention. Hence, attention will be focused on those four states that were, at the end of September 2009, leading in terms of installed wind power capacity. These includes California, but also Texas (in the lead), Iowa and Minnesota. Nevertheless, California, by dint of its considerable generation of electricity from geothermal and biomass as well as wind, is still roughly on a par with the generation of electricity from renewables (mostly wind) in Texas. California is also important as a key battleground over renewables incentives structures, and thus considerable attention is given to the state in this chapter.

I continue to pursue the two theoretical concerns of the last chapter. The first concern is the nature of ecological modernisation. Can we, in our consideration of more contemporary renewable policy and implementation developments, profitably operate a mode of analysis of EM that focuses on how the established (in this case electricity) industry has responded to environmental pressures? A different approach is one that is more oriented to that of what I have described as 'identity' EM. This sees the renewables industry as an independent force acting on the established electricity industry as the main path for development. Which is a more useful analysis?

The second theoretical question is: when we consider how outcomes occur, is it useful to put emphasis on the notion of 'interest' as a concept that can be independent of 'discourse'? In this chapter there is a continued focus on discussing the usefulness of analysis of the conceptions of both discourse and interests rather than focusing on one at the expense of the other. To continue the theme covered in the last chapter, while we may doubt the existence of objective notions of interest, this does not discount the identification and analysis of interests for the purposes of analysis of outcomes. To make a simple empirical illustration drawn from this chapter's discussion, renewable energy interest groups may adopt a discourse that they favour market-based trading mechanisms, but it is difficult to understand outcomes without assuming that in practice their interests rely on guarantees of long-term incentives to do business.

Wind power capacity in the USA has jumped more than tenfold since 2000. At that time most of the capacity was installed in California, but by 2009 they had been overtaken (in wind capacity) by Texas and Iowa, with Minnesota coming in at fourth place in terms of installed capacity. Together these four states accounted for just over half total US wind power capacity in the summer of 2009. Hence I focus on these four states in answering the empirical research questions and discussing the two theoretical themes. It may be useful to begin by describing some general aspects of the US renewables programme and then to look at these four states.

Renewable Portfolio Standards: A gap between discourse and practice?

The 'Renewable Portfolio Standard' (RPS) system is one of two key pillars that incentivise renewable energy in the USA. Its basic design is simple in that it mandates electricity suppliers to ensure that a given target proportion of electricity is supplied by renewable energy by a certain date. Penalties of various sorts can (in theory) be applied if this does not happen, or if the suppliers do not put in place plans to ensure that it should happen. A majority of US states now apply RPS of many different designs. There has been some thorough analysis of what may be best practice in RPS design (Langniss and Wiser 2003), but rather less attention to the institutional context that surrounds them. As indicated in Chapter 1, the notion of institution is drawn widely to include ideational, material, organisational and even technological factors.

A prominent feature of RPS politics is that they have been associated with markets in green electricity certificates or credits. Such things, at least in so far as they contributed to an image of market-based efficiency, have been a key part of the argument in favour of RPS institutions (Rader 2000). RPS is the system of choice of the American wind industry as represented by the American Wind Energy Association (AWEA). According to Rader, an RPS mechanism involving trading in renewable energy (certificate) credits:

> provides considerable flexibility to both retailers and renewable energy generators, which allows the market to meet the requirement more efficiently. That means minimizing the cost borne by consumers and thus bolstering political support for the policy. With tradable credits, retailers need not own or purchase renewable energy if they have no particular interest or expertise; they can simply purchase credits on a short- or long-term basis. Retailers who do get involved in renewable energy acquisitions can use credits to reduce risks: credits can be purchased to make up any production shortfalls or sold to take care of any excess. Renewable energy generators will find it easier to sell credits to multiple parties than to negotiate numerous power sales agreements. Another advantage of particular benefit to intermittent wind and solar energy generators is that credits can be sold at any time, regardless of when the power was generated. (Rader 2000, 403)

This discourse about achieving cost savings through the efficiency of the market by trading in specially devised renewable energy certificates (also in sulphur or carbon abatement emissions allowances) is common among environmentalists in the US, and has been led by The Environmental Defense Fund. They have espoused a dominant storyline about environmental governance in relation to pollution abatement schemes: '"Cap and trade" harnesses the forces of markets to achieve cost-effective environmental protection. Markets can achieve superior environmental protection by giving businesses both flexibility and a direct financial incentive to find faster, cheaper and more innovative ways to reduce pollution' (ED 2009, 1).

This discourse overlaps with the storyline about the operation of RPS through market-based mechanisms involving trade in green (renewable) electricity certificates (credits) (RECs). As will be argued in this chapter, the importance placed on the market mechanisms of the RPS system in the practice of renewable energy capacity expansion itself is misplaced. It will be argued that, while voluntary trading in renewable electricity certificates as part of green electricity schemes acts in the USA (in contrast to Europe) to expand renewable energy capacity, this should not be confused with the operation of RPS systems. The key point about the RPS mechanisms is that they work to get utilities to issue PPAs to developers rather than through the RPS–RECs. It may be in the interests of environmentalists and renewable energy generators to argue in the language and discourse of market trading in certificates leading to cost-effective outcomes, given the discursive dominance of market philosophy in the USA. However, we need to understand how the renewable incentive structure interacts with the perceived financial interests of renewable energy developers and generators in practice in order for us to gain an understanding of how the relatively rapid growth in renewables has been taking place. That is what I shall do, and it involves consideration of the interests of actors such as renewable developers and incumbent electricity utilities.

Besides the RPS system, a second key pillar of the incentive system for renewables in the USA is the production tax credit (PTC). The PTC was born in the period following what seems now to be the 'mini' oil price spike of 1991 and contemporaneous concern about the need to boost alternative energy supplies. PTC provisions were appended to a law passed in 1992 that was focused on promoting what was called deregulation in electricity. I say 'what was called' since deregulation in electricity does not mean a reduction in regulations; arguably it means more regulations, since it creates new markets which need to be defined and governance structures which need to be described. 'Electricity market liberalisation' is perhaps a slightly less misleading term.

The PTC has been extended, in a start and stop fashion, ever since. In legal and accountancy terms it works as a tax incentive. Persons and companies with tax liabilities can offset their investments in renewable energy against tax for a period of 10 years. What is important to the interests of the renewable energy developer and the generators is that this has the same effect as a subsidy of (in 2009 prices) 2.1 cents/KWh for a 10-year period. The key thing here is that it provides investors and banks with relatively long-term confidence in future returns. It is the provision of confidence by some or other means in minimum future levels of return that is important to capital-intensive investments such as renewable energy, and this is something that the

PTC achieves, albeit to the limited extent of its 2.1 cents/KWh price for only 10 years.

In 2009 the Obama administration introduced a grant incentive giving a similar benefit to the PTC. However, a fall in natural gas prices and thus a fall in the value of power purchase agreements on offer to renewable energy developers meant that this change did not stem the decline in the number of new wind projects that occurred in 2010.

The other necessary ingredients are contracts to supply energy, power purchase agreements (PPAs), usually agreed by the utilities. The fact that most renewable development occurs in states with RPS systems is associated with the tendency of RPS policies to pressure utilities to issue contracts for energy supply to renewable developers, something that will become clearer as the case studies are considered in this chapter. I argue that the inducement to offer long-term PPAs to renewable developers is the important effect of RPS systems, rather than their character as a device that involves trading in renewable energy credits.

This provision of confidence about future returns is the open secret behind the success of feed-in tariff systems in Europe. It has been the incentive regime of choice because it fits the interests of the majority of developers in Europe in countries such as Germany and also Denmark (until 2003). Spain also boasts a very successful feed-in tariff, although there the story is not as simple as is sometimes understood. However, in Spain, and all countries that boast big renewable energy deployment, the secret of success has been the provision of confidence in future returns. In the USA this has partly been achieved through the PTC in those states where there are other regulatory measures (especially RPS policies) in place, as we shall see later in the chapter. The PTC gives the same effect as a feed-in tariff in this sense of providing a significant period of confidence about future returns. This conclusion is obscured by some of the debate. One can recognise this conclusion while still sympathising with Gipe's (2003) criticisms of the PTC. These criticisms include its favouring big companies rather than community ownership, inducing opaque patterns of ownership, and being intermittent in nature (although this intermittency issue is an artefact of the Congressional Bills proposed rather than the instrument itself). Perhaps we ought to talk of the PTC as a 'rich man's feed-in tariff'. The point is that the interests of the corporate renewable energy developers (the most common type in the USA) are served by the PTC system, which acts in important senses for them like a feed-in tariff. From the point of view of maximising the volume of wind power capacity the biggest problem with the PTC itself, in addition to its intermittency, is that it is only a modest inducement. If the PTC were higher, then more wind power would be deployed. Indeed, in 2009 there were some moves to replace the tax credit aspect of the PTC with a straight subsidy (Broehl 2008). This would have provided the same level of incentive, but of a type that would have been more acceptable to the community wind power activists.

Of course, as will become clearer in the discussion about the different states, the existence of the PTC and favourable wind conditions are not the only factors that can account for the development of wind power in different states. The existence of favourable institutional conditions is also necessary, and, as already mentioned, the provision by electricity utilities to developers of contracts to supply renewable energy is a key aspect of such institutional arrangements. I shall place great importance on these factors in the discussions on the individual cases of the four states. The following sections of this chapter deal with renewable energy developments in the four leading new renewable states: California, Texas, Minnesota and Iowa. I begin or in this case continue (from the last chapter) with California.

California stalling

As was mentioned towards the end of the last chapter, the discourse in the 1990s was about competition and deregulation leading to cost savings, and attempts to give contracts to renewable energy developers were successfully opposed by the utilities on the grounds that it would disadvantage them as deregulation was implemented. As the California Energy Commission put it in 1997:

> Competition holds the potential for lower prices and more choices for energy consumers beyond those that can be obtained through continued monopoly provision and regulation of energy services ... Allowing competitive supply of energy products and services can lower the cost of basic services. It will also increase the number and variety of value-added options through technological innovation. To achieve these benefits, both state and federal governments must continue to facilitate the development of competitive markets by removing barriers. (California Energy Commission 1997, 10)

Hence renewables could only be promoted within a discourse of competition. In fact, this was already the path that had been observed, via a competitive bidding process for allocating contracts to renewable developers, which became the favoured policy mechanism at the end of the 1980s. This involved prospective developers putting forward bids to develop stated amounts of generating capacity for a price that they would be paid for each unit of energy produced. It is arguable whether much would, in fact, have come out of this process, as the accepted bids have been widely criticised as being too unrealistic to be viable (Asmus 2001, 162–3). Indeed, this is not the only time that problems have been experienced with competitive auctions of contracts. They have a tendency to produce low-cost contracts, which tend to be uneconomic and therefore not built, something that will be touched on later when we discuss more recent suggestions for competitive bidding schemes.

During the 1990s the attractions of so-called 'deregulation' dominated the energy agenda. Indeed, they took priority over pressures to implement 'demand-side management' (DSM) and renewable energy. The pro-renewables and environmental lobby was divided between different solutions for renewable energy. Some favoured a surcharge being attached to consumer bills which could then be used to fund energy efficiency and renewable energy, while others favoured giving the electricity suppliers obligations to supply prescribed proportions of their electricity from renewable energy. On the one hand, the American Wind Energy Association (AWEA), mainly representing independent developers, argued for an RPS system. This involved the utilities having an obligation to purchase renewable energy credits sufficient to meet their obligations (Rader and Norgaard 1996). The RPS proposal was backed by the Union of Concerned Scientists, and gained the support of the California Public Utilities Commission. On the other hand, EDF analysts, working with regulatory officers from the utilities, proposed a surcharge-funded 'production incentive' (Kirschner et al. 1997). Under this plan 'auctions' would be held in which renewable energy developers could bid for production incentives that would be awarded according to the lowest bids. The surcharge plan was eventually adopted (Wiser et al. 1998), but no renewable energy projects resulted. This was because the utilities would not offer any contracts that would enable them to be built (interview with Jan Hamrin, 15 April 2009). It is notable that both proposals talked about markets and competition, but the proposal from the AWEA was the one that actually gave the utilities, in effect, commitments to issue contracts for production (precisely what the struggle over SO#4 contracts had been about in the previous decade). The surcharge proposal did not achieve this.

There is a parallel in the contrast I have made earlier between two types of EM analysis. This is, first, in terms of analysing EM as a process whereby existing industry worked to improve its products in response to environmental pressures, as evidenced by the collaboration between EDF and the utilities. Second, we can use a different analysis of EM, which involves a focus on how renewable energy has grown as an independent force. In this case an analysis focused on renewable industries as independent forces seems relevant.

California experienced a crisis in 2000 as electricity prices spiralled upwards. The 'deregulated' electricity system was seen to have failed in its prime purpose, that of keeping prices low. The crisis was blamed on poor market design, including the decoupling of generation capacity-building from demand pressures (Woo 2001). Renewed arguments for an RPS utilised the post-crisis desire for means of building up capacity (Beck et al. 2002). Given that the previous 'surcharge-based' strategy had not induced the utilities to allow renewable development, there was more support among the pro-renewables lobby for the RPS solution, with, for example, the Sierra Club campaigning for adoption of the RPS proposal (Sierra Club California 2002). This time California enacted an RPS mandating the utilities to generate 20 per cent of their electricity from renewable energy by 2017. However, given that the state already generated 12 per cent of electricity from renewables, this was not such a radical target. After 2002 there was, until 2008, little renewable construction but considerable pressure (supported by Governor Schwarzenegger) to increase the RPS target so that 33 per cent of the state's electricity would be supplied by renewable energy by the year 2020.

Although the RPS system had first been put forward in California, other states took up the idea before California, and by the year 2000 some thirteen states had enacted some form of RPS (Berry and Jaccard 2001). Now some twenty-nine states have adopted some form of RPS and the mechanism has been widely embraced by environmental NGOs as a crucial means of promoting renewable development and one of the key pillars of the energy–environmental agenda. Carl Pope, Executive Director of the Sierra Club, in support of one the attempts to institute a nationally mandated system of RPS, said:

> The two biggest steps the House (of Representatives) could make – improving vehicle efficiency (CAFE) standards, and establishing a national market in every state for renewable electricity (RPS, or Renewable Portfolio Standards) are still being fiercely resisted by a reactionary coalition of American auto manufacturers, southeastern public utilities, and the coal industry. (Pope 2007)

It should, however, be borne in mind that the RPS system is something that was first taken up by AWEA and has been steadily promoted by them during the course of a campaign in which AWEA gained more consistent support from their natural allies in the environmental movement. The design of the California RPS is, however, markedly different from that in other states (Wiser et al. 2005). In particular it does not involve trade in renewable energy credits (although, as will be discussed later, it is dubious whether this really makes much difference). A problem identified by 'stakeholder analysis' (Wiser et al. 2005) is that there were perceived to be inadequate measures to deal with transmission upgrades necessary to connect wind power schemes in particular. Jan Hamrin described this problem (which is generic across the USA and beyond for wind power):

> the process for approving new transmission was not designed for renewables, it was designed for big central station plants. To be eligible for a transmission allowance you had to have the plant ready to go with the permit and everything needed for the plant and then you could apply for the needed transmission. If you are doing small modular types of projects like renewables, and they are being developed by the private sector they just did not fit smoothly into that process. You had a circular problem – you could not get a contract until you got a transmission allocation but the utilities wouldn't sign a contract until you had a transmission allocation. (Interview with Jan Hamrin, 15 April 2009)

As will be discussed in the case of Texas (which seems to have overcome the transmission problem), there is a sense that transmission authorities are reluctant to pass on big bills to consumers for building the transmission infrastructure necessary to support wind power. It seems that only sustained and well-supported campaigns for the transmission-building programmes will overcome this lack of perceived interest in building transmission upgrades (interview with renewable energy expert in Austin, 1 May 2009). There has been some progress in developing transmission capability to develop wind farms in California. Action by the Federal Energy Regulatory Commission (FERC) to encourage the development of transmission capacity necessary for wind power expansion is beginning to have an effect in California, with new lines being implemented. It appears that the amount of wind power generated in California is once more subject to considerable growth, although in the meantime two other states, Texas and Iowa, have overtaken California and others, such as Minnesota and Washington, are now not far behind.

Pressure for the transmission upgrade action necessary to achieve sustained development of wind power to achieve major proportions of electricity supply requires altering the electricity infrastructure. However, once again it has required political pressure, on a national basis, to upgrade the transmission infrastructure to achieve this, rather than any decision by the utilities themselves, which have tended to proffer transmission inadequacies as an explanation for slow development of renewables. The AWEA was active in developing and supporting efforts by FERC to ensure that transmission upgrades are achieved to facilitate wind power deployment (AWEA 2007). Hence, once again, ecological modernisation in the case of renewable energy cannot be seen only as the way in which industry reacts to pressure, but rather as the way in which a new renewables industry acting as a force (with its own identity) independent of the state-based electricity regimes (the utilities) has pressured the main electricity industry via the state to conform to its developmental needs. There is little doubt that the renewables lobby, working in a coalition with environmental groups, sends out a distinctive policy message, which it presses home to encourage (with greater or lesser success) reluctant electricity utilities to adopt. The pattern of companies that develop renewables is more complex.

However, a gap has opened up in some people's minds between the public efforts of the main electricity utilities to meet the ambitious RPS targets and the realities of achieving those targets. There is a fear that the utilities are issuing contracts to renewable energy developers that may not be developed, and hence that the RPS targets will not be met. The system being used by utilities involves a 'request for proposal' (RFP) system, which then involves contracts being issued to the lowest bidders according to criteria set by the utilities. A problem with such competitive bidding systems is that companies can make low bids to secure contracts, but then it is often found that the projects cannot be financed because they are judged to be uneconomic or too risky. NGOs are complaining that the utilities will not be sufficiently accountable for shortfalls in RPS requirements and may not be penalised for gaps between the volumes of contracts issued and the amount of renewable electricity that will be generated. One NGO said, in an appeal to the President of the PUC:

> Recent reports indicate that California's principal IOU's, Pacific Gas and Electric, Southern California Edison and SEMPRA, are struggling to meet the 20% RPS by 2010 mandated by law...Clearly, it is in the public interest that the CPUC's rules for determining an IOU's (investor owned utilities) compliance be readily accessible and transparent. The Commission should not allow lax compliance of our state's RPS laws and regulations. Discretionary waivers would betray

California's fight against global warming; and fossil fuel dependence for electricity generation. (Gonzalez and Freeman 2009)

In fact, there are clear signs that large amounts of new capacity are coming on line, including significant quantities of wind power, geothermal, biomass energy and solar plant. Most of these projects are being organised by companies that are independent of the California utilities, although some will have connections with 'new energy' wings of established utilities in other US states or other countries, and some will certainly be bought up by such interests later. FPL Energy, for example, is a subsidiary of Florida's main electricity utility, and now owns around 800 MW of wind power in California (Business Wire 2006). A lot of the solar thermal proposals are associated with companies based as far away as Israel and Spain. However, some doubts have been cast over the viability over some of the projects that have been given PPAs, most of all over some proposed solar thermal plant. This leads us on to a discussion of the politics of solar power, both large-scale and small-scale, in California, in part because accusations are being made that some of the large solar projects being proposed are unlikely to materialise. Again we are faced with the need to consider how discourse and interests must be considered as separate concepts in order to achieve a good understanding of policy outcomes.

California Solar?

Much has been said about the prospects for solar power in California, although some argue that there is a gap between discourse and practice. We are talking about two types of solar power here, solar photovoltaics (pv) and solar thermal electric: first, solar pv.

Since the beginning of 2007 there has been a California Solar Initiative, which aims to install around 3000 MW of solar pv in California, mainly in residential settings. The programme's biggest incentive involves a state-sponsored subsidy of $2500 per kW of installed capacity. Although this sounds a big subsidy, in fact it will cover only around a quarter of the cost of an installation. According to a California State rule introduced in 1996, solar pv and other renewables can offset the consumption of the residence against the cost of buying the same units from the supplier though what is called 'net metering'. This produces an incentive structure that is more favourable to solar pv in the case of rich consumers in California. This is because of the electricity tariff structure employed in California. One of the successes of the environmental lobbyists in the 1970s and 1980s was to reverse the incentives offered by utilities, which used to be to offer low unit prices for bulk energy purchases. Instead, residential consumers who use large amounts of electricity (in particular to air condition large houses, sometime called 'McMansions') have to pay much higher marginal unit charges for their electricity consumption – approaching 40 cents/kWh. This gives these consumers a much greater incentive to invest in solar pv than the usually much less wealthy consumers who live in smaller houses.

Installation is running at a rate of around 150 MW a year (North Carolina Solar Energy Association 2009). This is a significant market, yet only around a tenth the size of recent annual capacity additions in Germany. Critics of the current programme argue that a similar outcome to Germany needs a similar type of support instrument, that is, a feed-in tariff. Certainly, at current installation rates, the 3000 MW solar pv target would take over 15 years to achieve. Even this may well be unrealistic under the terms of the current programme, because the subsidies are set to be reduced as the capacity installed is increased. Supporters of the programme are criticised for being too optimistic about the decline in costs of solar pv (interview with Paul Gipe, 6 May 2009). In 2008 the capital costs of solar panels fell after an excess of panels, some from China, came onto the market, which may inflate expectations of the rate of decline in capital costs. It is certainly the case that previous opinions about the speed with which solar pv can become competitive were over-optimistic. David Elliot, writing in 1997, said, for example, that 'Within a decade or so they (solar pv cells) are likely to be competitive with conventional power sources' (Elliott 1997, 87).

'Vote Solar', the most visible lobbying organisation representing US solar pv installation interests, supports a 'reverse auction' system for awarding contracts to install solar pv in California. This idea, favoured by the California PUC, in fact is broadly similar to the system used by the utilities to award contracts for larger projects. Although sometimes called a type of feed-in tariff, it is not recognised as such by supporters of feed-in tariffs, since these involve fixed prices being offered rather than auctions being held. However, critics of this proposal fear that this could lead to many projects not being implemented, in much the same way as the larger projects. Nevertheless, according to Vote Solar, the system will be better than the 'feed-in' tariffs used in Europe:

> Some governments have used standard-offer, fixed price feed-in tariffs to incentivize renewable energy development. The difficulty with this approach is finding the right price. If the price is set too low, it does not stimulate the desired market activity. If the price is set too high, ratepayers pay unnecessary costs, suppliers throughout the value chain are not encouraged to reduce prices, and the program can lose political support. By using a market mechanism to determine the contract price, the CPUC's program uses competition to establish a price that is both sufficient for project development and protective of ratepayers. With the price of solar modules coming down 40% over the past 6 months, we expect dramatic market activity at price levels that will attract the interest of policymakers around the country. (Vote Solar 2009)

The discourse of market competition is again being used, but it is not possible to understand the position of Vote Solar, and companies such as SunEdison and SunPower, who support their position without understanding their interests. These interests are defined by the incentive structure underpinned by the electricity tariff structure operating in California, which gives companies selling pv systems the possibility of a market selling to wealthy owners of large, high-energy-consuming houses. Solar pv companies do not have the same interests under different tariff structures used elsewhere.

Viewed in this way, the distinction between state interventionism in the market to set prices and leaving this up to market competition largely disappears. Such market as there is in California for solar pv rests on the basis of three different state interventions in the market, which set an essentially fixed price for electricity from solar pv. There is the net metering rule, the $2500 subsidy scheme under the California Solar Initiative and the residential electricity tariff structure designed to encourage conservation by big energy users. As also discussed in Chapter 1, analysis based on institutional economics (Hodgson 1998) would recognise such a description. When we speak of 'markets' we do not speak of some sacred zone where individuals and firms operate in an area that is free from predispositions and formal or informal rules that might influence outcomes. The appeal by 'Vote Solar' to ideology, that of the discourse of free market economics, can be seen as an attempt to justify their interests in terms of a dominant free market discourse in the USA, and California in particular. It is therefore useful, contra Hajer (1995), to discuss the interests of the solar pv industry, of giving consumers calculable, fairly precise incentives, as prior assumed 'facts' to the discourses that are put forward as a justification for their policy positions.

Solar thermal electric, as mentioned in the last chapter, is one of the renewable technologies that were implemented in the 1980s as a result of the California renewables programme. The technology was the 'solar trough' system whereby mirrors concentrate light onto a fluid running along the centre of the mirror troughs. The fluid is passed to a central steam turbine generator. This is one of the technologies being given power purchase agreements, although other solar thermal technologies are also being given contracts that are less well tested. Indeed, the capacity of the contracts being issued to solar thermal projects has brought criticism from the wind power industry that speculative solar thermal projects are being given contracts in preference to wind power schemes. It is claimed that at least some of the solar thermal projects are unlikely to be built because the prices that the developers bid to be paid under the PPAs granted are unrealistically low (Broehl 2009, 29–30). According to Jan Hamrin:

> The utilities have signed huge, speculative contracts for 5282 MW of new solar thermal projects, the majority of which have never before been built on a commercial scale. The trough projects have been built commercially (California has 354 MW of solar trough generation that have been producing power for more than 10 years) but new trough plants are only 1540 MW of the total. There are three large power tower projects proposed and 1750 MW of sterling dish projects. There has never been a commercial sterling dish project built to date. These contracts allow the utilities to say that (1) they have fulfilled their RPS obligation by signing contracts; and (2) there is a good chance a lot of these are never going to be built. To reinforce my cynical view, PG&E announced...it had signed a contract for a solar satellite project. (Interview with Jan Hamrin, 15 April 2009)

The solar satellite project, organised by Solaren, would involve reflecting solar energy from space and sending it back down to Earth. Certainly if the solar power projects are being proposed then the developers have an interest in developing them. However, there are also other possible interests at play. It has already been stated that the utilities have an interest in signing PPAs that would fill the ambitious RPS mandates they have been set regardless of whether the projects are implemented. While the developers may fully wish to develop their projects, they still can benefit from winning the contracts even if the projects are not carried out. Companies can raise money through equity offers. They can also sell on the contracts to other companies. Jan Hamrin argues that the scales are tilted by the utilities in favour of solar thermal and against wind power when analysis is done using the 'market reference price' (the price used by the Californian regulatory system to measure what energy supply would cost to provide without the renewables being considered):

> In California wind does not cost more than the Market Reference Price (MRF) and has not resulted in higher electricity rates. RPS Solar projects however have contracts that are significantly higher than the MRP and will result in the need for public goods funds (PGF) to be used to make up the difference. Moreover, most of the solar contracts signed in the last few years in California have been the result of bilateral negotiations rather than having been won through the competitive bidding process that all wind projects must go through. (Personal communication from Jan Hamrin to the author, 21 October 2009)

Many people are willing to give the proponents of what some cast as speculative solar projects the benefit of the doubt. After all, the untried machines that benefitted from the tax breaks to be installed in the 1980s in the 'wind rush' were given the benefit of the doubt, and, while many machines failed, the experience did pave the way for what is now a multinationally based (wind) industry. However, there is also a feeling among renewable advocates that some utilities may not be so keen on wind power as an option (interview with Mike Sloan, 25 April 2009; interview with Jan Hamrin, 15 April 2009). For a start, the utilities can earn an assured return on investments in traditional power plant, these days usually new combined cycle gas plant, while wind power in California is seen as relatively expensive and also is seen as a 'non-despatchable', intermittent supply by many people in the electricity regime establishment.

On the other hand, solar thermal electric plants, while far from cheap themselves, are more comparable to traditional power plants. On this point Hamrin says:

> Solar thermal plants look a lot like utility steam plants, so it appeals to utility engineers. They are not wimpy looking like wind or biomass

plants they look like 'manly' central station generation plants. You have got a lot of power being produced in one central spot. The utility can control it. You also have some new designs that now incorporate storage with solar trough as well as power tower technologies that will allow these plants to be dispatchable. With the trough technologies they have the capacity for 6 hours of storage, which means that gives them a lot of leeway to match peak loads which makes the resource much more valuable. With the power tower I believe they can have as much as 16 hours of storage capability. (Interview with Jan Hamrin 15 April 2009)

Texas

I want to set out some key objectives for this section. One objective, following on from the introductory section of this Chapter, is to move away from discourses about the relative market efficacy of the Texan RPS and focus more on how institutional conditions have existed in Texas to give renewable energy (mostly wind power) developers the confidence in achieving long- term returns on their capital investments in wind power. Again it is an issue of considering how interests are served rather than merely about how discourses become dominant. Another objective is to continue the theme of analysing EM to discuss the extent to which the thesis that renewable energy has been an independent industrial and policy force (compared to the conventional electricity industry) in Texas. Texas is important to study simply because it has the largest amount of installed wind power compared to any other state in the US. By October 2009 installation had reached around 9000 MW which will produce around 7 per cent of electricity consumed in Texas.

Discourses

One discourse that is repeated is that of how a key role of Texas is to lead the nation's charge towards 'energy independence'. As the Governor of Texas says: 'Texas is better at producing energy than anyone else in the United States, which is why we need to lead the charge on energy independence.' Texas is a leader in crude oil, refining and natural gas production, and wind energy production, but the governor said Texas must continue to expand alternative energy technologies, including nuclear, solar and clean coal (Office of the Governor Rick Perry 2008).

Also, 'Texas is an energy state, it is a "can do" entrepreneurial state' (interview with renewable energy expert in Austin, Texas, 1 May 2009). On the other hand, another analyst says:

> ...because wind power for many years has been cost effective here (in Texas)...because of the way it is structured with the federal incentives. If you get to a windy area and you are competing against expensive marginal fuel costs which we have here, Texas is sort of the perfect storm of events for wind power...in Texas, we have had an enormous amount of growth (in electricity demand) and we need more power....Texas more so than anywhere else in the country, is one of the top in the world in structuring competitive energy markets for electricity. They are recognised as having the most competitive wholesale electric market. (Interview with Mike Sloan, 25 April 2009)

There is also an academic discourse about the role of the RPS in achieving wind power development in Texas. Langniss and Wiser (2003, 534) comment: '[I]t can be said with near certainty that, given previous development plans, the major driver in the resurgence of wind energy development in Texas has been the state's aggressive RPS.'

I would certainly say that the RPS was associated with persuading the utilities and electricity companies that wind power deployment was a good business that needed to be pursued. It gave effect to policy demands made by the widely supported and well-organised lobby for renewables in Texas. However, I am less certain about the academic focus on the details of the RPS itself as opposed to other regulatory issues. Much academic discussion about the role of the RPS focuses on the details of RPS design, but this focus may be misplaced.

In particular, I would argue that the actual design features of the Texan RPS are not any better, as such, than those of the California RPS. Langniss and Wiser (2003, 528–9, 533) pay considerable attention to the details of RPS structure. Wiser et al. (2005) imply that big problems are associated with the design of the Californian RPS. They recommend changes to procurement practices, contract failure, bundling of Renewable Energy Credits (RECs) and transmission reinforcement and upgrade issues. This may be helpful, but it does not explain how it is that there is much more wind power being built in Texas. The Texan RPS itself contains very few details, and its targets are smaller than the RPS targets in California. The point that needs to be made here is that, if the comparison is between California and Texas, the much superior outcome in terms of Texan deployment of wind power probably has little to do with the different RPS design features in Texas and California. Certainly, transmission upgrade planning policy appears to be going more smoothly in Texas than in California, but even here, this has little or nothing to do with the design of the RPS itself, but rather other institutional and legislative arrangements.

Another factor is the degree of planning opposition to wind power. There are much fewer cases of planning opposition to wind power in Texas compared with California, where some places, such as the California Bay area, are virtually 'no go' areas for wind power, but this, again, has nothing to do with the design for the RPS. Bohn and Lant (2009) find a strong association between the complexity of wind power planning laws and the amount of wind power deployed, with higher deployment being associated with less complexity. An explanation for this is simply that in places like California, where there are relatively greater controversies associated with wind power planning applications, there are likely to be more complex regulations adopted by governmental institutions in order to mediate such controversies.

Another factor may be that in California the utilities are simply less inclined to favour wind power compared with other, arguably more expensive, solar thermal electric technologies. This has been discussed earlier. In Texas many wind power projects have been financed with contracts paying between just under $30 and $60/MWh. In California relatively fewer proposed projects have been within this range in recent years (interview with Mike Sloan, 25 April 2009; Wiser and Bolinger 2008, 17). Partly this is about Texas having considerable wind resources compared with California, but it is also about policy choices made by electricity utilities.

Until recently Texas has had little facility for giving contracts to non-wind power renewables, including solar and biomass. There may be justifiable concerns about the practicality of some of the proposed solar thermal projects in California, but at least some are likely to be going ahead, while the equivalent plans for solar thermal electric plant in Texas are much less well developed. California seems much more likely to achieve a significant near-term increase in capacity of solar thermal and other non-wind renewables than Texas. This suggests that what restricts wind power development in California compared with Texas has not much to do with any inferiority of California's RPS design (*per se*) compared with RPS rules in Texas, but it may have something to do with prevalent attitudes among electricity utilities.

Hurlbut (2008, 159) implies that RPS details are not the only determinant of renewable outcomes. In particular, he says: 'Details matter, but perhaps the most important and applicable lesson arising from the Texas experience is that it also matters how the RPS fits into the big picture.' I would go even farther than this in the sense of arguing that, in comparing the Texas RPS design with the California RPS, consideration of the RPS design details does not matter very much at all.

The discourse about the workings and necessities of having markets in Renewable Energy Credits (RECs) to make RPS systems work may seem important because of the dominant storyline in Texas about the importance of markets, but it is difficult to see what role the RECs have had other than as an accounting device to measure target compliance. Discussions about whether RECs should be bankable or not, the mechanisms for enforceability of the penalty system, and the systems of trading RECs are certainly now irrelevant as far as the continuing rapid growth of wind power is concerned. That is because wind power development, now around 9000 MW, has greatly overshot the mandated portfolio requirement of 5880 MW. The RECs are now effectively worthless.

Nevertheless, the evidence is that voluntary green power markets, or green electricity trading schemes, have had a big effect in expanding the market for renewable energy. This happens especially in states where there is no RPS system. A report published by the National Renewable Energy Laboratory commented: 'Green power markets provide an additional revenue stream for renewable energy projects, and raise consumer awareness of the benefits of renewable energy' (Bird et al. 2009, 32). Jan Hamrin comments:

> The voluntary (green electricity) market allows developers to build their RPS projects a bit larger and sell excess RECs into the voluntary market as supplemental revenue (the power is sold into the wholesale market). It also allows RE developers to bring their projects on line earlier than their RPS contract might specify and sell the output into the voluntary REC & wholesale markets until the utility needs the power for RPS purposes. And this supplementary revenue from the sale of RECs into the voluntary market is critical for states without an RPS. (Jan Hamrin in personal communication with author, 21 October 2009)

This conclusion may surprising to analysts of green electricity markets in the UK and other EU countries, where it is very unclear what they add to the incentive structures put in place by the state. Such schemes have been criticised because it is so often the case that the certificates are being 'double counted'. This is because, for instance in the UK, at least in the past, the certificates may be subsidised by all electricity consumers, with the 'green energy' consumers having to pay little or nothing extra for the electricity. However, in the USA there have been, from an early stage, clear legal restrictions on green electricity schemes which prevent them from 'double counting' the incentives that come from all consumers, such as Renewable Obligation Certificates (ROCs) in the UK.

This argument about green power markets should not be confused with the role of certificate trading in the Texas RPS. The establishment of the RPS was important for the development of wind power in Texas, but its own significance has nothing to do with the efficacy of market trading in RECs. Rather, it has to do with the RPS being an institution that induced the electricity regime in Texas to accommodate the new renewable industry, at least in the form of the wind industry. Again, we need to consider how interests and institutions influence outcomes.

Institutions and interests

There was a pale reflection of the battles over power station construction in the 1990s as the state Public Utilities Commission experienced a battle between utility plans (by the company TXU) to build new gas and coal-fired power plants and NGOs' attempts to stop them. In doing so, the NGO lobby embarrassed the energy establishment by pointing out clashes of interest between state energy decision-making and the state's natural gas production industry. Although Texan oil production has dropped to a small proportion of what it was, natural gas production has risen to make Texas by far the largest natural gas producer in the USA. Companies such as Occidental Oil and the Texas Industrial Electric Consumers have provided opposition to the renewables lobby. However, the rise in gas production is offset in its material consequences for the balance interest groups' pressures in Texas by the fact that electricity consumption is rising very fast in the state, having risen by 45 per cent in the 1990–2007 period (EIA 2009). Texas has easily the highest electricity consumption of any US state, and the rate of increase gives room for new wind industry market entrants.

As has been implied already, an important aspect of the Texas renewable energy strategy is the relative ease with which it is planning the transmission upgrades needed to bring wind power from areas such as

West Texas and the Texas Panhandle into the cities. 'We have a wholly intra state electric reliability region called ERCOT, the Electric Reliability Council of Texas, meaning we are not federally regulated on our transmission, within the ERCOT region. It is very important, it is like being an electrical island and means the rules can be managed with relatively fewer regulatory authorities' (interview with renewable energy expert from Austin, Texas, 1 May 2009). Upgrades sufficient for 18,000 MW of wind power have been agreed and are being constructed. As discussed earlier, wind power needs, in contrast to conventional power stations, transmission upgrades to be installed in advance, and this has been problematic in places like California where FERC needs to coordinate actions between different authorities if transmission lines are to be agreed in the context of transmission zones that span several states.

In the early 1990s an NGO campaign on power plant approvals was led by Public Citizen, a liberal Nader-type citizen campaigning body, and Karl Rabago featured prominently in this campaign. A relatively liberal Governor, Anne Richards then appointed Rabago to be one of the Public Utility Commissioners. Rabago was later replaced by a more establishment figure, Pat Wood, and went on to work for EDF and to argue for the renewables lobby.

In the lead-up to the 'deregulation' restructuring of electricity in 1999, it was agreed that a 'deliberative poll' as advocated by Fishkin (1995) be held to test opinion on renewables. The result, which surprised the utilities, was that renewables were a favoured option and that consumers would support a small extra cost to pay for them (interview with renewable energy expert from Austin 1 May 2009). This result boosted the campaign by the pro-renewables coalition for what was now (by the end of the 1990s) the renewable lobby's campaign for an RPS. The 'Wind Coalition' was the strongest part of the pro-renewables lobby, and key supporters of this coalition included FPL and independent companies such as Orion (from California) and Renewable Energy Systems (RES) (from the UK). In Texas FPL Energy was independent of TXU, which dominated the Texan electricity supply. Also involved in the Wind Coalition were some elements from TXU itself who were involved in development of demonstration wind farms. Alongside this lobby group, an organisation called 'Texas Renewable Energy Industries Association' (TREIA) was significant.

When electricity restructuring occurred, in 1999, this was a little later than in California. This allowed Texas to become more or less the first case of an RPS design being put into legislation during what Hurlbut (2008) highlights as being an example of a Kingdon-style 'policy window' (Kingdon 1984) for the benefit of renewable energy interests. TXU wanted to offload onto consumers the stranded costs of power plant investment in order to improve its market position to face competition on the wholesale electricity market that was being created in the so-called 'de-regulation' process. TXU needed public consent for this, and they gained this by bargaining with the renewables lobby to achieve their support for a restructuring deal. Both interests found that a political deal meant a positive sum outcome for them both. This involved the establishment of an RPS with a target of 2000 MW, initially by 2009. However:

> The dirty little secret is that probably the (renewable) advocates settled for far less than they should have, now knowing what the utilities were thinking and how ready they were to do it. In fact, they gave away billions in terms of stranded cost adjustments for old power plants in return for a relatively small investment in renewables. (Interview with renewable energy expert in Austin, Texas, 1 May 2009)

The key change made by the RPS, and the policy debates that surrounded its establishment, was to create an institution whereby the electricity suppliers began to issue PPAs to wind power developers. The fact that wind power was economically competitive made this an easy thing to do, and the Texan electricity 'regime' learned to invest in wind power. Various independent (from TXU as the dominant utility) entered the market, and TXU itself invested in wind power. In 2005 the mandated target was increased to 5880 MW by 2015. An indicative target of 10,000 MW by 2025 also exists, but that does not currently have any compliance consequences or financial incentives attached. The issuing of contracts to wind developers became as routine as the issuing of PPAs to other independent developers. This is the key development that flowed from the inception of the RPS.

Included in the 5880 MW target by 2015 is a target of 500 MW for new non-wind projects, although by March 2009 around 200 MW (mostly biomass) had been built. It is difficult to build non-wind technologies, especially solar power, since it will have difficulty competing in the prices available for PPAs in the current market structure. A lobby has been working in the State Legislature to secure a firm RPS in favour of other renewables with a proposal for an RPS for 1500 MW of non-wind renewables. Implicitly this would involve utilities offering PPAs to renewable developers for higher rates than are currently on offer to wind developers. The lobby pressing this case included TREIA, Public Citizen, EDF, Environment Texas, Texas Business for Clean Air and the Sierra Club, with Bright Source Energy (a solar thermal electric company formerly called 'Luz') being the only stated energy company involved (EDF 2009).

In surveying the way that interests interact with institutions to produce favourable outcomes for wind power in Texas, it is conventional to see planning rules and traditions of permissiveness for development in Texas as being generally conducive for wind power. Planning decisions are made at County level and there are usually few objections to wind farm proposals.

Minnesota

Texas and California may make up a high proportion of new renewable energy capacity (that is, apart from large hydro), but it is important to do at least an outline check on other important states to see whether the story we are picking up so far is consistent.

Minnesota is one of those states where wind development began in the 1990s, and its position as a state with few coal resources and also a strong 'clean energy' lobby has led it to be one of the early leaders in developing wind capacity. By the middle of 2009 there was around 1800 MW of wind power installed, as well as just over 300 MW of biomass generation capacity. This has to be put in the context of Minnesota's small size. Its electricity consumption is only a fifth the size of California's and around a seventh the size of Texas'.

The state began setting renewable energy goals in the 1990s, and in 2003 an indicative target was given to the utilities to acquire 10 per cent of their electricity from renewables by 2010. In fact, this has now been achieved, but in 2007 the state officially adopted an RPS mandate requiring that the utilities deliver 15 per cent of their electricity from renewable sources by the end of 2010, rising to 30 per cent by 2020. The state does seem on track to achieve this. Initially the wind ownership has been mainly by independents such as Iberdrola, Enxco and Horizon Wind, but the dominant utility, Xcel, is now taking an interest and is developing wind farms to own for itself.

However, once again, the success of the RPS system does not appear to lie in any system of trading renewable energy credits. Indeed, the means by which the Minnesota PUC would enforce compliance are vague, especially given that the utility programme for renewables is described as a 'good faith' effort. The institution that will keep the utilities compliant can perhaps be described as one involving bargaining rather than one resting on financial incentives. Excel is the leading electricity utility in Minnesota, and has latterly been investing in wind power. As a leading Minnesotan wind analyst puts it:

> They (Xcel) have a whole long list of things as a regulated industry that they want to accomplish in a state and if they are totally ignoring the renewable energy standards or the renewable energy goals that are set, they are going to have a more difficult time getting what they want. I would say they have been gritting their teeth about this policy, it was not their idea to do it, that is why it was mandated, why it was set in standards. (Interview with Lisa Daniels, 11 May 2009)

Rather as in the case of Texas at the time of restructuring, the pro-renewables lobby and allied environmental groups have political assets of public support and they will only loan them to the utilities if they achieve some basic aims of the renewables lobby. In this way there is a policy network existing in this state that resembles a stasis somewhere in between the 'policy' community and the 'issue network' described by Marsh and Rhodes (1992). In a policy community there is a high degree of consensus and resource exchanges between the actors, while in an issue network there is much disagreement and few resource exchanges. In this case we can see that some resources are exchanged and some agreement is reached, although the ideal position of both parties remains unsatisfied. This situation is to be compared with California, whose renewable policy network most closely resembles an 'issue' network. However, part of the explanation for this is that in the cases of Minnesota and Texas there are good wind resources. Prospective wind power developers in Minnesota can therefore claim contracts from the utilities without costing them very much, if anything, in terms of increasing the prices they have to charge their consumers. This is much less the case in California. See my earlier comment that different access by wind power actors to political resources is based on differing 'material' institutions, which are in turn based on differing wind resources.

In Minnesota there is a strong 'community renewables' lobby. Minnesota has a strong agricultural cooperative tradition through the dairy industry and other industries. The cooperative tradition has more recently manifested itself also in growing crops for ethanol production. In the 1990s the renewables lobby had a strong agricultural basis.

> Minnesotans for an energy efficient economy had a pretty broad based coalition started, there was also the sustainable energy for economic development which was where we combined clean energy advocates and the agricultural and rural interests. That was very successful, together those organisations and the Isaac Walton League and Clean Energy, Clean Water Action was also a strong. (Interview with Lisa Daniels, 11 May 2009).

By 2009 there was around 430 MW of 'community'-owned wind power, that is, mainly by farmer cooperatives, and also capacity owned by municipal organisations. Hence over 20 per cent of wind capacity in Minnesota is community-owned. This has been promoted by an organisation called 'Windustry', and the proportion is the highest in the USA. According to Wiser and Bolinger (2008, 15), in the US as a whole in 2007 just 2 per cent of wind power capacity was community-owned, 84 per cent was owned by independent power producers and 14 per cent by incumbent utilities.

Iowa

Iowa's percentage of electricity derived from wind power is, in October 2009, the largest in the USA, with 15 per cent of Iowa power coming from 3000 MW worth of installed capacity. Iowa can claim to be the first state to have initiated an RPS. A commitment to providing 2 per cent of Iowa's electricity from renewables became an effective policy, despite initial opposition from the utilities, by the beginning of the 1990s.According to Ed Woolsey (interview 27 May 2009), it was a populist campaign that drove the demand for a renewable RPS: 'The populism came from the ownership aspect, smaller farms and small co-ops, actually being able to provide for their own energy and actually the ownership option there of being able to afford to enter into this new energy industry.'

However, by 2000 support for renewables was increasing. The Iowa environmental council and the Iowa renewable energy association were important in lobbying activity. The utilities were under legislative pressure to give way to this lobby. According to Ed Woolsey (interview 27 May 2009):

> [The Governor] Tom Vilsack, he called in his chairman of the utility board [the state PUC], which was his appointee, the chair of the consumer advocates office, the CEOs of a couple of utility companies and said basically... 'we want to have more wind in Iowa, what would it take to do that, to get you on board?' They said, 'we have got to be able to make money off it'. This is my [EW] conjecture, having played a part in it and seeing the results. They [the utilities] said 'we have to make money', the [the Governor] said 'what kind of policies would it take?' They wrote the policies, he advocated it to the next legislative session and basically it gives the utilities 11.77 to 12.43% return on every dollar they put into wind and transmission and anything associated with wind, so basically give them a good rate of return on their investment into wind.

This having been done, the utilities started to invest in wind farms. MidAmerican Energy is the largest (by sales) electricity utility in Iowa and has installed the largest amount of wind power, over 1000 MW by the summer of 2009.

It should be said here that the pro-renewables lobby that won the important institutional breakthrough was rather different from that in Texas, and more similar to the rural, farmer-oriented lobby that brought in policies favouring wind power in Minnesota. As in Iowa, there is a significant farmer ownership of wind power, although the lion's share is now being deployed by mainstream developers. Ed Woolsey (interview 27 May 2009) argues that more community ownership is needed to encourage greater acceptance for large-scale wind power. He points to planning resistance beginning to emerge in some places and is arguing that the long-distance interstate transmission network that is being built to improve wind power grid connections needs to incorporate more local ramps to enable farmer-owned wind power to be more easily connected at a more local level.

Once again, in institutional terms the focus has been not so much the size of the RPS or its design rules, but the provision of important profit-making incentives to the utilities. Iowa, like Minnesota, has not been subject to electricity 'deregulation', and the utility interest in wind power stems from negotiated regulations on what the utilities will contract for and under what conditions they will invest in renewable energy plant and grid infrastructure. In Iowa there are no renewable energy credits to be traded. Hence, just as in the other three states that have been studied here, incentives or trade in renewable energy credits has little or nothing to do with promoting the installation of renewable energy.

Conclusion

Two theoretical questions were posited in the early part of this chapter, first about ecological modernisation analysis of the electricity and renewables industries and secondly about the use of discourse and interest-based analysis. On the issue of ecological modernisation analysis, it is clearly more useful to analyse this case as the emergence of a renewables lobby, with its own identity, which is often challenging the existing electricity industry. The existing utilities, who still make up only a relatively small minority of renewable developments, have seen their interests, at least initially, as being opposed to changing practices significantly to accommodate renewable development. This may be partly because incumbent utilities are influenced by fossil fuel and nuclear power generation interests to preserve their generation interests, and it may also be because the engineers and managers have learned routines and practices that pay little regard to renewable energy. It is also a matter of incentive structures. Another aspect of 'identity' EM is that separate incentives are given to renewables, and these exist in the form of RPS at state level and the PTC at Federal level. These institutions are complemented by 'green power' markets that rest directly on consumer identification and support for renewable energy – a significant type of identification, since the green power supplies will, on average, cost significantly more than conventional electricity supplies (Bird et al. 2009).

Ecological modernisation theory needs to be understood as involving key 'identity EM' characteristics in the case of renewables development in the US. This includes the renewables lobby acting as a force largely independent of the established electricity industry in terms of policy bargaining. It is correct to point out that, on closer inspection, these independents turn out, more often than not, to be owned by big energy companies themselves. Certainly FPL Energy (now known as NextEra Energy Resources) is owned by the incumbent Florida utility FPL. Since 2000 a lot of the independent renewable developers have been bought out by renewable energy development wings of multinational companies. EnXo has been bought by the 'new energy' company owned by EDF Energies Nouvelles, in turn owned by the French giant EDF. Orion has been bought up by BP through its 'Alternative Energy' company. Horizon has been bought up by the Portuguese electricity major, EDP, again through its renewables company. However, even after these takeovers the renewables lobby cannot be regarded as a subset (at least in public policy terms) of the electricity giants. As will become clearer when considering other cases, companies operating in particular contexts operate according to the incentive structures in those contexts, usually with surprisingly little regard to the policies of their parent companies. Iberdrola Renewables is owned by the Spanish multinational electricity company Iberdrola, although, as we shall see later, even in Spain the Renewables wing has a noticeably different agenda from the electricity major that owns it. FPL may not do enough to favour renewables in Florida, but its 'new' energy company NextEra Energy Resources (ex FPL Energy) will battle with incumbent utilities in other states to secure conditions that are more favourable to renewable developments.

Hence, once again we can see that, even though the renewables industry has become more mainstream in terms of its ownership, Mol's theory (1995) of how ecological modernisation can be analysed by examining how the established industry has responded to NGO demands for environmental improvements is still too simplistic for this renewables case study. So, if we are going to develop ecological modernisation theory to account for how a new and very distinctive ecological technology emerges, we should not simply treat it as another example of how an existing industrial establishment attunes itself to environmental lobbyists. In this case the leading lobbyists trying to change the practices of the electricity establishment are the renewables lobby themselves. They act in alliance with environmental and other NGOs to gain concessions from the existing fossil–nuclear–electricity-distribution-and-supply establishment. Initially they were outsiders to this system, and to a large extent they still are in a policy sense in the USA when tackling the utilities.

Hence, besides the identity EM characteristics mentioned earlier of a distinctive incentives structure for renewables in the four states and the existence of an independent renewables trade lobby, we can also see two other characteristics of identity EM in the form of coalitions between renewable energy interests with environmental NGOs (often arguing with utility interests) and also the deployment of the projects being largely organised by companies at least different from those of the incumbent utilities. The last identity EM characteristic is less clear than the other three, since sometimes the renewable development companies can be arms of incumbents in other states (notably FPL).

The glue that binds the pro-renewables coalition together is the notion of an alternative, new, renewable technological identity. It is this alternative identity that drives support for incentives for renewable energy technologies, partly as a reaction to the perceived dirty and resource-depleting identity of the conventional electricity generation technologies. In this sense we can see an illustration of how, even in a case (the contemporary US renewables sector) where there is so much involvement of the mainstream electricity companies, what has been called an 'identity ecological modernisation' approach is well placed to encapsulate development.

This case study also illustrates why it is important to analyse interests rather than rely on discourse analysis. If utilities work under rules that allow them to earn money out of building gas-fired power stations, but not renewables, then they will not develop renewable schemes themselves and may not cooperate in terms of offering contracts to independent renewable developers. If the utilities are in a situation, as in California, where many wind power proposals require extra price support that is considered to lead to increases in rate charges for electricity consumers, the utilities may be unwilling, again, to offer a large amount of contracted capacity to independent developers.

Of course, it is possible to overcome such problems. In Iowa the rules were changed to allow utilities to make profits out of renewable development directly. In Texas and Minnesota the utilities have been persuaded through bargaining to give power purchase agreements to independents, since the utilities want regulatory conditions to favour them in areas other than renewables policy. The fact that renewable deployment is a popular objective means that the support of the renewables lobby is an asset that the incumbent utilities (or dominant electricity companies in liberalised markets) are willing to 'buy' in order to secure various objectives.

We can see, during the course of this analysis, how important it is to discuss how the actors pursue their interests – and also how these interests are perceived and the contexts that lead to such constructions of interests. We can also see how these interests attempt to appeal to dominant discourses, such as the discourses that emphasise markets, efficiency, trading and competition. However, as with the American Wind Energy Association and its espousal of RPS, the practice of meeting these interests is not necessarily achieved through acting out the discourses themselves. Trading in renewable energy credits seems to have little to do with the way in which RPS policies advantage renewables. In short, RPS systems work on the basis of traditional practices of regulatory bargaining, so that renewable energy developers are given power purchase agreements that simply guarantee them the value of wholesale electricity for extended periods. The RPS paraphernalia of renewable energy credits is little more than window dressing to convince people that a fairly conventional approach to electricity regulation is really some sort of free market trading device. Contrary to Hajer (1995), it seems that in this case interests cloak their preferred practices in 'free market' discourses rather than the discourses conditioning the practice of the renewable developers.

On the other hand, the dominant free market discourse in the US does seem to reduce the possibilities for making interventions to help renewable energy, or at least opponents of giving renewables greater price support seem to be able to mobilise this discourse to that end. At the time of writing there is around 30 GW of wind power in the USA, enough to supply just over 2 per cent of US electricity, but that is still a much smaller percentage than in a number of EU states. As will be discussed in the next chapter, on Europe, 'feed-in tariff' rates for wind power and other renewables are much higher in many countries compared with the USA. To some in the USA this means that the subsidy system in the EU is less efficient. To others it just means that the US is not willing to put more effort into deploying renewables.

Indeed, as US wind deployment slowed in 2010 compared with 2009 in the wake of lower natural gas prices, the US renewables programme could be characterised as being based on the development of a modest amount of capacity provided that natural gas prices stay high. Certainly some analysts believe that the feed-in tariff system should be extended to the USA. According to Sovacool (2008, 201–3), a federal-based system of feed-in tariffs, set at different levels for different renewable energy technologies, should be introduced. He argues that such a move is necessary to create a rapidly growing set of markets for renewable energy technologies, that it would reduce electricity prices, and also that 'a national FIT would create harmonisation, consistency and predictability for financers, investors, manufacturers and producers' (Sovacool 2008, 203). Despite the growth in deployment of renewables, so far mainly wind power in the USA, there is quite a way to go before the quantity of renewable energy developed begins to make way for an absolute reduction in the use of non-renewable energy resources. In the 1997–2008 period there was an 18 per cent growth in electricity demand, while in

2008, for example, renewable electricity expanded by only around 0.5 per cent (US EIA 2010). Certainly an acceleration of renewable deployment, and also much stronger efforts to reduce demand-side pressures, will be needed to significantly reduce the absolute quantity of non-renewable energy being used in the electricity sector alone.

6

Germany, Spain, the UK, Australia and China

This chapter looks at renewable energy development in selected countries with the objective of studying the politics behind the inception and growth of renewable energy and classifying them according to key analytical criteria that are of special focus in this book. A key research question is: 'What association is there between the appearance of characteristics of "identity" ecological modernisation and different levels and types of renewable development in the different case studies?' In order to remind the reader of what was said in Chapter 2, the 'identity' EM approach posits five characteristics in relation to the renewable energy case: idealism in the technological innovation process; the existence of an effective dedicated financial support mechanism for renewable energy; the existence of independent renewable energy trade associations; the formation of coalitions between the renewable trade association(s) and environmental NGOs; and deployment of renewables by companies that are independent of the main energy companies.

The key outcome is the amount of renewable energy deployed in each country, the main types of renewable energy deployed, and the relative speed with which this has been achieved. The analysis will be facilitated by dividing the analysis into a discussion of the birth and development of the renewables industry in each case, an analysis of the interplay between discourse and interests, and also a discussion of how far the different cases may be characterised as being 'identity' EM.

Some of the early parts of the EM story on renewable energy have already been told. Chapter 3 discussed the invention of the modern forms of wind power and biogas technology. Chapter 4 discussed the role of California in acting as a handmaiden of the process of transforming a nascent wind power industry. This chapter covers how this process of creating an industry was replicated outside the US, focusing mainly, although not exclusively, on wind power. Denmark, Germany and Spain were the earlier runners in developing significant volumes of wind power capacity, and, indeed, until 2008 these three countries still had the largest capacities and percentage shares of electricity consumption from wind power in Europe. Since 2008 other EU states have overtaken Denmark's capacity (it is a small country), but not its percentage share of electricity from wind power and new renewables in general (around 30 per cent).

The Danish case has been widely discussed, and also summarised in Chapter 3. Expansion of onshore wind power in Denmark has stalled on account of the policies of a conservative coalition, so this chapter concentrates on cases where there has been considerable recent expansion of renewables. The cases are selected to represent a range of different case studies and conditions in order to examine how, and to what extent, renewable energy has developed as a separate identity and how discourse and interests have interacted in this process. Germany and Spain are selected for particular focus precisely because they have been early leaders in developing the renewable energy industry. If we want to make an effective analysis of ecological modernisation processes in renewable energy, then it is necessary to pay some attention to these leading cases. A key focus here is the question of whether a greater or lesser association with 'identity' EM as discussed in Chapter 2 is associated with the degree of development of the renewable energy programme in a particular country.

It is also necessary to make other comparisons in order to acquit such analysis and put it in an understandable context. Hence other country-case studies are done; cases that have been slower in taking up a determined policy to promote renewable energy. Might mainstream EM be more relevant in such cases? By 'mainstream EM' I mean (as discussed in detail in Chapter 2) the expectation (implicit in Mol's 1995 study of the chemical industry) that the choice of technologies to be implemented in pursuit of environmental objectives is made by the main energy companies rather than through public pressure. The UK and Australia are selected for study here as examples of cases where large renewable programmes have been started later compared with Denmark, Germany and Spain. These two Anglo-Saxon countries have had less of a tradition of anti-nuclear and grass roots energy action than the latter three states.

Finally there is an outline study of China. This book has focused on Western cases; yet there is a need to analyse what is happening in developing countries to understand the extent that EM (identity EM

Table 6.1 Percentages of electricity available from new renewable sources in five cases in 2009

Germany	12
Spain	12
UK	5
Australia	1.5
China (wind power only)	1.5

Sources: Office of the Renewable Energy Regulator (2009), Directorate General for Energy and Transport (2008), Böhme, D. (ed) (2009), AEA Energy and Environment (2009), interview with Hugo Lucas (IDAE) 9 November 2009; estimate from 20 GWe of installed wind power in China assuming 22 per cent capacity factor.

or mainstream) might be suitable also for analysis beyond the Western world. Although it has not been possible to devote (in this work) sufficient resources to undertake a full case study, sufficient work has been done to allow some tentative conclusions to be drawn about the extent to which identity EM analysis could be extended to the case of China.

First there is a comparison of the current sizes of the renewable energy programmes, as in Table 6.1. As can be seen, Spain and Germany have the largest programmes, followed by the UK, with Australia in fourth place.

Germany

Birth and development of renewables programme

Germany was the scene of some of the most severe confrontations over nuclear power. Many instances of demonstrations, sometimes violent, and prolonged political action by coalitions of anti-nuclear groups, backed by a range of civil organisations, characterised mounting opposition to nuclear power construction (Rudig 1990, 149–66). The anti-nuclear movement began to gather serious momentum in 1975, and by the end of the 1980s further nuclear power construction had become politically impossible. Locally based 'Citizen's initiative' groups sprouted up in the 1970s. They pressed for action on environmental issues. In the 1980s such groups often became principally focused on anti-nuclear issues (Markovits and Gorski 1993, 102–4). This became associated with a desire to engage in energy initiatives that provided an alternative to nuclear power. As one activist put it:

> traditionally Germany had a very strong anti nuclear movement and in the early 70s and 80s more and more people said 'no' to nuclear but they had no 'yes'. 'Yes' to what? They wanted to have

electricity. So there was a need for the anti nuclear movement to develop solutions...they copied the first wind turbine machines from Denmark in the '80s...It was a grassroots technology in the 80s and the beginning of the 90s. This grass roots movement was asking for feed in tariffs...asking for the right to connect these machines to the grid and the big utilities refused to do it...(until) after almost 10 years of lobbying...it was mainly farmers and environmentalists in the north of Germany and they just connected their turbines illegally to the grid. They just fed it in. (Interview with Sven Teske, Energy Campaigner for Greenpeace Germany, 14 September 2004)

This grass roots pressure led to a series of attempts by backbench members of the Bundestag to propose Bills to effect a feed-in law. Such was the pressure from below that even the conservative CDU/CSU came to support the idea. The first feed-in law came into effect in 1991 and it has been maintained, refined and reinforced ever since so that it covers a range of renewable energy technologies. The technologies received fixed prices for electricity sold to the grid, guaranteed over 20 years. The legislation was achieved and maintained in the teeth of opposition from the energy utilities, and also the initial opposition of the Ministry for Economic Affairs (Jacobsson and Lauber 2005, 133–5). In fact, even in the conservative south, there was support for feed-in tariff legislation, since there biomass and hydro were being used as energy sources by landowners and farmers (interview with Sven Teske, 14 September 2004). As was discussed in the last chapter, key initiators of farm-based biogas technology came from Southern Germany.

Renewable energy became a consensus political objective, and major renewable energy industries were developed. As the 1990s wore on, the rate of renewable energy deployment, especially that of wind power, accelerated. By the middle of 2009 Germany had established sufficient renewable energy capacity from sources developed under the feed-in laws to supply 14 per cent of German electricity; just over half of that was from wind power and much of the rest from various sorts of biomass sources. Biogas sources are expanding in output, and there are moves to use the biogas directly as a gas fuel in its final form by sending it through the natural gas grid.

Two points need to be made here with respect to central themes of the book. The first concerns how, in a very independent manner, the new renewable alternative was pursued with its own identity, which contrasted sharply not only with nuclear power but also with fossil fuels. At this time pressure for feed-in tariff laws was mounting. Public concern about global warming was also rising. Indeed, in the early 1990s concern about global warming was greater than in all other comparable Western countries (Brechin 2003, 110). This public concern was reflected at the government level by a strongly worded report by the Enquete Commission in 1989–90, including a commitment to reduce carbon dioxide emissions by 30 per cent by 2005. Such commitments were not to be adopted by, for example, the UK until around 20 years later. In fact, the radical carbon dioxide reduction targets were prompted partly by a drive from pro-nuclear advocates among Christian Democrats to provide a justification for nuclear power in the face of strong anti-nuclear pressures (Hatch 2007, 43). This proved unsuccessful, but the radical commitments on climate change sharpened the dilemma facing social movements and policymakers who were also against nuclear power (since the policy objectives conflicted). The strong programme favouring renewable energy can be interpreted as a response to that dilemma.

An expansion of nuclear power would be the response to global warming that would have favoured the interests of the electricity industry in Germany as they then existed; the strategy of mainstream ecological modernisation would therefore suggest this, although EM theorists have, paradoxically, traditionally opposed nuclear power. So the mainstream EM position of seeing ecological problems solved by action that is organised by and internalised by the mainstream conventional industry does not account for the rise of renewable energy in Germany. Indeed, it raises more questions than it answers. The utilities association, the VDEW, actually lodged a complaint with the European Commission to try to stop the feed-in tariff laws. Ultimately this stratagem failed to have an impact on the feed-in tariff legislation because of the broad cross-party support that the legislation enjoyed (Jacobsson and Lauber 2005, 135–6).

Discourse and interests

Here I want to continue my discussion of how (contra Hajer 1995) it can be that material interest can be analysed as being independent of discourse in the case of incentives for renewables. A key point argued here is that the interests of renewable energy developers for long-term power purchase agreements in the US is much the same as in Germany, even though the discourses appear to be conflicting.

The discourse favouring feed-in tariffs (articulated by independent renewable energy generators in Germany) is hotly contested by the main electricity suppliers/generators in Germany, such as RWE and

E.ON. However, their position has shifted over the years from outright opposition to the notion of special incentive schemes for renewable energy in the 1990s towards support for a tradable certificate scheme (Stenzel and Frenzel 2008, 2650). They have made small forays into the renewables market. It is anticipated that they will make large investments in Germany's offshore wind power programme. There have been controversies about grid access for independent companies.

Eurelectric is the European peak association for the main electricity companies. Although some of their members, especially the Spanish, actually favour feed-in tariffs, the majority still do not. They tend to articulate a position that is close to that of the German electricity utilities. Eurelectric is very much in favour of promoting the notion of tradable green certificates and is critical of the German feed-in tariff system. A spokesman for Eurelectric said: 'to promise a new wind generator now an income for 20 years from this day, we just do not think that is appropriate in a market. Also if technology is not competitive in the next 5 to 6 years, why should we subsidise it?' (interview with Juho Lipponen, 19 December 2007).

The Eurelectric spokesman further commented on the practicalities of striving to achieve the target of 20 per cent of energy from renewable energy by 2020, which was the key aspect of the 2009 EU Renewable Directive:

> Our fear is we feel if we really want to reach the ambitious level of 20% we will need more something like a third of the electricity generation but if we reach it by supporting renewable energy outside the market so it is not really part of the money base, it is not part of the market, it is taken to the grid and you don't have to sell it so we would end up possibly walking away or losing one third of our electricity market we are trying to develop. (Interview with Juho Lipponen, 19 December 2007)

Eurelectric did not succeed in achieving its aim of making trading in green electricity certificates a part of the EU Renewables Directive. However, it does seem apparent from this quote that the idea of creating markets in green electricity certificates that are associated with the achievement of renewable energy targets fits in with the interests of the major electricity companies. The major electricity companies say that this system is more cost-effective, although, as will be seen in later discussion about the British and Swedish systems, this is contradicted by analysis of the performance of these systems. What the green electricity certificate does do is give control of the renewables market to the major electricity suppliers, since only they can offer power purchase agreements that are likely to be regarded as creditworthy by banks and investors in the renewable energy schemes. The battle over incentive structures for renewables is often, and certainly is in this case, a battle for market share among different interest groups.

The perceived self-interests of German electricity utilities should be considered in the context that a) the large bulk of electricity supply comes from either coal or nuclear power. b) In the context of an electricity market that has only grown very slowly since German reunification in 1990, renewables can be seen as a threat to the profits of the coal and nuclear power stations.

The interest of the independent, grass roots generators were associated with a need to be paid rates for the electricity they produced at levels that were guaranteed in the long term. They needed this because of the way the schemes were financed. The developers raised money from banks, as well as equity contributions from local farmers or the public through share offers. However, banks would only lend money to the projects if there were guarantees of future income flows. The state-owned bank *Kreditanstalt für Wiederaufbau* lends 80 per cent of the capital at low interest rates (interview with Ralf Bischof, BWE, 13 September 2004; Lauber and Metz 2004, 604). It may be argued that a 'discourse' about the rightness of feed-in tariffs emerged. There is certainly a literature espousing their effectiveness and importance for promoting renewable energy programmes. 'The majority of countries seeking a policy to rapidly increase their renewable energy capacity employ the FIT model. It has driven the deployment of more renewable capacity, and at lower cost, than any other' (Mendonca et al. 2010, 9). The point is that it exists because it is an institutional arrangement that suits the interests of the renewable energy developers in Germany who have been the actors responsible for doing the lobbying that secured the passage of the feed-in legislation.

Renewable energy developers in the USA apparently favour 'market-based' incentives, whereas German renewable interests want fixed prices for renewable income streams set by Government. Yet, despite these discursive differences, the priorities for both German and American renewable interests are the same; they need long-term power purchase agreements.

In Chapters 4 and 5, covering the USA, it was discussed how the independent companies needed power purchase agreements from the utilities (in effect a private form of feed-in tariff). Nevertheless, wind interests in the USA put forward a discourse concerning how trading in green electricity certificates was an important part of Renewable Portfolio Standards (RPS) in the individual states. It became clear in the analysis that the discourse of tradable certificates served mainly as a means of gaining political acceptability for the notions of RPS – the renewable companies still needed power purchase contracts which were achieved through the RPS, and the certificates are more or less irrelevant to the working of the RPS. The results in Germany and the USA have more similarities than is usually realised. In particular, in both countries, the major electricity companies are regulated to allow independent renewable energy developers access to electricity markets. Hence, I argued that interests dictated discursive tactics rather than discourse constructing the interests of the renewable interest groups. Where the cases differ is that the incentive structure for renewables is weaker in the USA than in Germany.

Identity EM

The German case exhibits an especially sharp contrast between 'mainstream' EM and the 'identity' EM described in this book. Not only did (and do) the renewable technologies have an identity independent of that of the conventional technologies (mainly coal and nuclear) that dominated electricity supply, but the development and ownership had an independent identity. Almost all of the wind power projects in Germany have been developed by companies that are completely independent of the electricity utilities. Many of the projects are owned by local cooperatives, consisting of farmers or local investors, and most of the rest of the wind power capacity is owned by independent corporate companies. These companies, at least up until recently, raised their share capital from public share offers. Changes in tax laws have made this a less attractive option for ordinary investors, and other, more conventional, means of financing these wind power developments are being used.

A further change is that local opposition to wind farms has increased since the 1990s. In regions such as North Rhine–Westphalia attitudes were very favourable towards wind power, but movements against them have been organised and local authorities are imposing stricter restrictions on wind power development (Breukers 2006, 177–221).

Parallel to the movement for deployment of technologies such as wind farms and biogas was the movement for solar pv. At the same time as the feed-in tariff law was put into operation in 1991, the Federal Government launched the '1000 rooftops' scheme (Jacobsson and Lauber 2005, 136–8; Jacobsson et al. 2004, 16). The lack of a national feed-in tariff suitable to support solar pv motivated activists to campaign for cities, which often part-owned local electricity utilities, to argue for these city-based utilities to cover solar pv feed-in tariffs. This movement was spearheaded by Aachen, Hammelburg and Freising. This pressure from below boosted a campaign for a national solar pv tariff, which was achieved in 2000. The coalition that formed the basis of the campaign for a solar pv feed-in tariff consisted of a range of solar trade associations and environmental groups (Jacobsson and Lauber 2005; Jacobsson et al. 2004, 15–19).Initially the feed-in tariff rate was set at around 50 eurocents per KWh. Despite political attacks, a high solar pv rate remained in place for a decade, although the German Government has now begun to reduce these rates. Nevertheless, by the end of 2009 well over 5000 MW of solar pv was installed (Blau 2010), generating over 1 per cent of Germany's electricity. This installed capacity represented more than the combined totals of solar pv capacity installed in the rest of the world. This strong German identification with the technology is fundamental in nature and is based on the notion that the technology should be implemented where possible, rather than assessed on the basis of cost–benefit criteria. Clear differences can be seen between the US and German attitudes, which underscore the degree to which German opinion identifies with the need to implement solar power as a means of emphasising sustainability rather than development defining the path towards sustainability. As one analyst says:

> They (Americans) all look at it from the viewpoint of the utility companies because they are locked into this thinking that power has to come from utilities and that utilities are powerful, so we have to negotiate with them. The German approach was different, it was we as citizens and the politicians, run the country, we tell the businesses what to do, so we can change the market and we can tell the utilities that you just have to pass on these costs if you don't like them, if you do not want to absorb them. (Interview with Craig Morris, 29 October 2009)

If we look back at the characteristics of 'identity' ecological modernisation discussed in Chapter 2, we can see that Germany displays all of the five characteristics of 'identity' EM. There has been significant idealism and continued public campaigning in support of renewable incentives in Germany since the early 1980s. More recently, sustainability and development have been interpreted independently of utility notions of what renewable energy can be economically implemented.

This involves a different conception of the balance between sustainability and development than is apparent in US policy.

There is a very strong independent renewables sector – and also opposition between conventional utilities and the independent renewables sector. There are also strong trade associations for each renewable energy technology, such as the Bundesverband WindEnergie (BWE) for wind energy. There have been some crucial coalitions between environmental groups and the renewable trade associations. There are strong renewable financial incentives, and these incentives have been won and maintained after major public campaigns and grass roots lobbying of Parliament. Altogether, the German case displays very clear contours of what we might call 'identity EM'.

Spain

Birth and development of renewables programme

Spain had in common with Germany, and also with Denmark (as we saw in Chapter 3), a strong anti-nuclear movement in the 1970s and 1980s. In Spain's case this militancy included confrontation and also became mixed up in anti-Francoist and nationalist struggles in the Basque and Catalan regions. This included what was hailed as the biggest ever anti-nuclear demonstration, against the Lemoniz power station proposal, and even as a justification for an explosion set off by ETA (Rudig 1990, 136–9, 212–16). The nuclear power programme was abandoned in the early 1980s and has never been revived. Anti-nuclear feeling was widespread, and motivated governmental and regional politicians to support demonstration renewables projects (interview with Josep Puig, 13 October 2009). Anti-nuclear sentiment was strong on the left in the socialist and communist parties, and the first socialist government under Gonzalez stopped the nuclear power programme after it was elected in 1982 (Lancaster 1989, 158–79). This formed key elements of the context in which left-wing activity, such as formation of worker cooperatives, was associated with anti-nuclear sentiments.

The anti-nuclear atmosphere was a background to the inception of the Ecotechnia workers' cooperative in 1981. The cooperative went on to design and manufacture dozens of 10–20 kW wind machines. Ecotechnia was formed by 'a group of nine persons with high technical qualifications, committed to environmental thought and the practice of alternative technology' (Puig 2009, 191). They were strongly influenced by writers such as Lovins (1977) and the practical experience of developing wind power in Denmark through initiatives such as the

Tvind project (mentioned in Chapter 3) (Puig 2009, 191–2). According to Josep Puig:

> a group of these people (the founders of Ecotechnia) were involved in the anti nuclear movement and many times when they were speaking against nuclear a lot of people asked that 'if we don't build nukes what to do?'...and at the time, some people knew that in Denmark they were starting with wind and they were developing quite well, so we decided to propose to build a wind machine of 15 kW with the government. (Interview with Josep Puig, 13 October 2009)

The Spanish Government established IDAE, the Institute for Energy Saving and Diversification, in 1984, and IDAE became a crucial, indeed, until the mid-1990s, the crucial, institution promoting development of renewable energy in Spain. According to Puig, one of the founder members of Ecotechnia, in the 1980s:

> Ecotechnia only sold 20 to 30 small machines, 15 or 30 kW to the different administrations in different regions of Spain because they wanted to make some demonstration programmes...IDAE started developing renewable energy programs...at that time it was a woman quite sensitive to renewables and anti nuclear in the ministry and also because the first director of IDAE was a man quite interested in renewables. (Interview with Josep Puig, 13 October 2009)

In the 1980s and for much of the 1990s the state was a crucial driver of the wind power programme, through IDAE and regional governments' investments in renewable energy, in what can be described as public–private partnerships (PPPs). IDAE was the key driver in the early days, but in the 1990s the regional governments took over the public aspect of the PPPs from IDAE (Dinica 2008). In 1994, the government of the region of Navarra co-financed the construction of a series of wind farms in collaboration with the electricity utility Iberdrola. One of Iberdrola's subsidiaries, Gamesa (initially a manufacturer of automotive parts), became the supplier of the bulk of the wind generators installed in Spain (Stenzel and Frenzel 2008, 2651).

A feed-in tariff was introduced in 1991, but this gave no long-term guarantees about the level of the feed-in tariff. Although a more stable version of the feed-in tariff was introduced in 1998 (giving a 5-year guarantee period), the 20-year guarantee was only introduced in 2004. Dinica (2008) argues that too much attention has been focused on the technicalities of the feed-in tariff mechanism to explain how confidence was built in the renewable energy programme compared with looking at the confidence-building activities of the state.

Certainly, by 2004, the Spanish wind power programme had taken off, and by 2009 constituted around 12 per cent of Spanish electricity production (interview with Hugo Lucas, IDAE, 9 November 2009). The major electricity utilities in Spain have led the development, with Iberdrola taking an especially prominent role through its subsidiary, Iberdrola Renewables. There is an agreement with the Spanish transmission operator, Red Electrica, to develop 40 GW of wind power by 2025. Additional quantities of renewable capacity will be developed through biomass, small hydro and solar pv technologies.

The solar pv programme is backed by a feed-in tariff at the same sort of level as the German feed-in tariff. This resulted in a massive increase in the take-up of the technology in the year 2007. Most of this has been through large schemes, something that is a little different from Germany, where household schemes have made up most of the projects and nearly half the total capacity installed (interview with Craig Morris, 29 October 2009). By the end of 2008 some 3000 MW of solar had been installed, compared with around 5000 MW in Germany (Fouquet 2009). There has been some conflict with the major electricity suppliers in that they have pressured the Spanish Government to restrict the programme because they are under an agreement with the Spanish Government to keep down electricity prices. In 2008 the solar pv programme was suspended, and after consultation a cap was imposed on development at a rate of around 500 MW a year (interviews with Josep Puig, Eurosolar, 13 October 2009 and Hugo Lucas, IDAE, 9 November 2009). High feed-in tariff rates are also being paid for solar thermal electric schemes and a range of other technologies, including various types of biomass.

Third, the apparent pattern of implementation by major electricity utilities is more complex than it first appears. In fact, Iberdrola, for example, operates its renewable business through a subsidiary, Iberdrola Renewables. Acciona, another major electricity company with investments in renewables, owns Energia Hidroelectrica de Navarra (EHN), which has major wind farm holdings. The renewable development companies owned by major electricity suppliers do not necessarily have the same interests as their parent companies on all issues. Indeed, the renewable development subsidiaries have sometimes put forward different positions compared with their parent companies. One example is the wish of the electricity companies to keep down electricity prices, which can conflict with demands for high feed-in tariffs. Another example concerns pumped storage capacity, which the renewable developers want to be the subject of centralised investment, whereas the parent companies are not so keen on this. They may want to continue making money by using their own pumped storage drawn from nuclear power (interviews with Josep Puig, Eurosolar, 13 October 2009 and Hugo Lucas, IDAE, 9 November 2009).

Discourse and interests

Spain is reliant on imports for the large bulk of its energy sources. The Spanish Government defines the importance of renewable energy as being central to the aim of reducing Spain's energy dependence:

> Spain has experienced a rapid growth in energy consumption. Our growing and excessive dependency on imports to supply about 80 per cent of our energy in recent years and the necessity to preserve the environment obliges us to formulate a sustainable means to ensure efficient use of energy and the utilisation of clean resources. So it is important to greatly increase the use of renewable energy and this is connected to the achievement of economic, social and environmental objectives. (IDAE 2005, 5)

The fact that the electricity utilities are not bedded down with domestic energy interests (there is little coal mining or oil or gas extraction) means that they can be favourable towards renewable energy, especially when investment in the other non-fossil fuel, nuclear power, has been stopped. State support for renewables from IDAE and regional governments (called 'autonomous communities') meant that the major electricity utilities could take advantage of this support and enter a new sector in an expanding electricity market. It has to be remembered that, in comparison to Germany, not only are there much fewer domestic energy interests in Spain to compete with renewables, but there has been a rapid increase in electricity consumption in Spain compared with relative stagnation in Germany. This all coincides to produce a context where it is much more in the interests of the Spanish electricity industry to promote renewable energy development than it is for the German electricity industry.

Although the initial work on development of Spanish wind power was spearheaded by a quasi-idealistic anti-nuclear workers' cooperative in the form of Ecotechnia, there has been little local ownership of wind power. The context for this is that the Spanish countryside is generally very poor, and farmers are poor, in comparison to Germany, where the countryside is much more affluent and there are many more possibilities for locally based energy activism and ownership of wind projects.

Instead, local municipalities tend to look to wind power development as a source of income via tax revenues, which is usually a major boost to them given their small size and low resource base (interview with Alfonso Cano, APPA, 9 December 2004). Far from opposing central government policy mechanisms to favour renewable energy development, the regions have tended to push for more of them in order to improve local possibilities for development. Spanish regions have tried, and very often succeeded, in linking wind power developments with contracts that stipulate that work is carried out in the locality of the developments. *Wind Power Monthly* commented in 2004 that: 'Spain's autonomous regional governments...see local wealth in the wind. In Navarra alone its 700 MW has created 4000 jobs, while regions like Galicia, Castile & Leon and Valencia insist on local assembly and manufacture of turbines and components before granting development concessions' (McGovern 2004).

The fact that there is much less opposition of interests between local interests and the major electricity suppliers means that there is no basis for a general anti-utility discourse. On the other hand, compared with Germany, there is relatively less planning opposition to wind farms. In an earlier analysis by the author and others it was said:

> In Spain, around one in five applications receives significant opposition from local wildlife conservation groups, and these can be subject to considerable delays. However, even these cases tend to result in more money being paid for local projects rather than municipal rejection of the schemes. In only some rare cases have municipalities simply refused to allow wind farms to be built. (Toke et al. 2008, 1134)

Again, the depressed economic environment in much of the Spanish countryside may be a factor here in explaining the lesser rate of planning controversy compared with countries such as the UK and even Germany. Low property values and a middle-class preference to go to the cities or the coast lessen the possibilities of anti-wind farm activism.

Dinica (2008) highlights a gap between academic discourse about the intrinsic benefits of feed-in tariffs and the interests of the renewable energy generators. The renewable energy support mechanism has evolved into a feed-in tariff system that is broadly comparable with the German system (although the feed-in tariff for wind power has an optional, 'market-based' element). However, the key factor that has made the Spanish renewables programme work is the confidence created in investments. As argued earlier, this confidence was achieved by governmental investments in the projects. This persuaded project partners, usually companies owned by major electricity suppliers, that they would achieve a reasonable rate of return. Once again, this example illustrates that often analysis should be aimed at understanding how interests influence discourse rather than just the other way around. Renewable energy developers do not have financial interests of a certain type because they support a discourse about feed-in tariffs, or green electricity certificates. Rather, they tend to propound discourses about the 'efficiency' and 'cost-effectiveness' of different renewable energy support mechanisms based on their interest in securing policy mechanisms that assure long-term financial confidence (for investors) in the renewable energy projects.

A consequence of this is that Spanish electricity utilities have diverged from their counterparts in Germany, in that in Spain they support the feed-in tariff system. This has evolved in Spain as a system that rewards the major electricity suppliers in an atmosphere where governmental intervention to secure rapid economic development has been an accepted norm. The Spanish electricity companies have firmly, and consistently, resisted the German utility (and Eurelectric) preferences for pan-EU trading in green electricity certificates, since this was seen as upsetting the Spanish utilities' interests in renewable energy. Claude Turmes, an MEP who has had considerable influence on energy and environmental issues, talked about the battles over the EU Commission's policy on tradable green electricity certificates in the 2001 EU Directive on renewable energy:

As on all European legislation, there is a big amount of lobbying around, so Eurectric was, of course, present to try and win the certificate model, whereas as the renewables industry was much more in favour of feed-in tariff... We won the battle because of the Spanish renewable industry being linked very much to the conservatives politicians in Spain. De Palacio who was the (Energy) Commissioner was very close to these conservative, very rich families which are running the renewable business in Spain and that was something I discovered in 1999–2000 when I went to Spain to conferences. I was used to see mueslis around the table on renewables, and in Spain I saw women with very rich jewellery and so we won the Directive at the end politically because the Spanish conservatives in parliament voted in favour of the feed-in system, or at least the possibility to have the choice between the two systems and the Spanish Commissioner, also conservative, was on that side. (Interview with Claude Turmes MEP, 27 February 2007)

Identity EM

Spain certainly exhibits 'identity' EM in terms of the important action by idealistically (anti-nuclear, left-wing) oriented early initiatives from Ecotechnia, supported by sympathetic politicians. On the other hand, there are elements of mainstream EM, in that the electricity utilities are the main developers of renewable energy. However, there are also several incongruities with mainstream EM. First and foremost, renewable energy is supported by environmentalists in clear preference to both nuclear power and fossil fuels. Renewable energy technologies draw their strength from being a fuel based on national consensus, but only because they are seen by many as preferable to nuclear power as a non-fossil fuel. Second, the initiation of the wind power programme in particular (which provides the large bulk of the new renewable energy) was achieved after activism in a practical sense (via Ecotechnia), which was based on activity well outside the main electricity companies and based partly on anti-nuclear and pro-environmentalist idealism.

Grass roots movements supporting solar power can be seen in action in the run-up to the adoption of Barcelona's 'solar ordinance'. This means that, from 2000, all new and refurbished buildings had to supply at least 60 per cent of their hot water needs through on-site solar power. The participation of grass roots actors in a coalition with renewable energy interests was evident:

> Before the final adoption of the Solar Ordinance in 1999, a long participation and negotiation process took place. In the process first phase, the main actors were individual local NGOs and a federation of NGOs such as 'Barcelona Estalvia Energia' ('Barcelona Saves Energy'), which was in contact with the Barcelona City Council to put energy efficiency measures and the promotion of renewable energies on the local agenda. Later on the Catalonian Association of Renewable Energy Professionals became involved. (Schaefer 2006, 4)

The campaign in Barcelona is especially significant since regulations requiring solar thermal panels to be installed throughout Spain followed in its wake. The idea of mandating the necessity of having on-site renewables also influenced policy in local authorities in other EU states by way of adopting regulations requiring developers to install on-site renewables in new buildings.

Spain is not as pure an example of 'identity' EM as the German case, because the main energy utilities are centrally involved in the deployment of the renewable energy programme. However, Spain's renewable energy programme still has key characteristics of 'identity' EM. Idealism, as represented by the inspiration for the formation of Ecotechnia, is something in the past, but it nevertheless formed an essential part of wind power development in Spain.

Indeed, we can see at least four out of five of the 'identity' EM characteristics evident in the Spanish case: significant idealism (in the early days) of the renewables programme: dedicated financial support mechanisms; independent and active trade associations; and coalitions with environmental groups. The coalitions with environmental groups are perhaps not so crucial in this Spanish case, although renewables enjoy solid support from groups such as Greenpeace. Nevertheless, they are important in operationalising the dominant Spanish public discourse in support of renewable energy as a crucial aspect of the nation's energy policy for minimising energy dependence. The fifth characteristic, that is, companies that are independent of the major electricity companies, is not apparent in the Spanish case, although, as discussed earlier, the interests of the renewable development arms of the electricity majors sometimes conflict with the interests of their parent companies. The Spanish renewable trade associations, however, do press renewable energy interests independently from the major utilities. The Asociación de Productores de Energías Renovables (APPA) acts as an umbrella organisation. However, in 2002, a specific wind association was launched, now known as the Asociación Empresarial Eólica. The Asociación Empresarial Fotovoltaica is the lead organisation representing solar pv interests.

United Kingdom

Birth and development of renewable energy programme

The United Kingdom can be distinguished immediately from the two preceding cases in that it had much less hands-on grass roots activity in developing renewable energy. The UK did have an anti-nuclear movement, but it was rarely associated with the sort of mass demonstrations, let alone confrontations, that occurred in Germany and Spain. Neither was there any parallel to either the workers' cooperative, Ecotechnia, in

Spain, which started making wind turbines in the early 1980s, or to the farmers and organisers of burgerwindparks in Germany, who connected wind turbines to the grid, also in the 1980s. Still less was there the sort of agricultural technology movement that invented wind and biogas technologies, described in Chapter 3 as being focused in Denmark. One factor here is that in the 1990s the UK enjoyed high levels of energy security in the electricity sector. However, this has declined as the UK's oil and gas reserves have been depleted.

There was considerable public interest in renewables, although until the 1990s this was resolved solely through R&D programmes rather than any commercial exploitation in the UK. In the 1970s this focused on wave power as the leading option. A cause celebre developed around wave power, focusing on an allegation that there was a sort of nuclear conspiracy to stop it being funded (Ross 1997). Stephen Salter promoted a well-known scheme called the 'Salter's Duck', which was intended to be placed in arrays away from the shore. His idea involved going straight from a design to very large projects very quickly. This looks, with the benefit of hindsight, over-optimistic given the problems that the more recent designs of wave machines have experienced on the road to commercial viability.

Changes in Government ideas about the priority given to wave power research may be put down to a mixture of mistakes and a general wish to cut public spending, and also a general lack of priority being given to renewable innovation.

Whereas, in the 1970s, wave power was regarded as the most attractive renewable option and wind power was well down the list, by the mid-1980s the order had been reversed (Watt 1998, 62–8). Perhaps we can explain this by the existence of a favourable dominant discourse on wave power, which created a general disposition to want to believe in the energy source. Talk about the waves, after all, fitted in with British traditions of surviving through mastery of the sea. On the other hand, by the mid-1980s there was evidence from Denmark that wind power was a more immediately practical proposition. By then the R&D programme had decided to spend some money on prototype wind turbines. However, the British response focused most resources on building big wind turbines, even though the engineers who were organising the projects felt this was a bad move. One, David Lindley, who later became the Managing Director of National Wind Power and the Wind Energy Group, commented:

The CEGB was being dragged along and their argument being that it was rather pointless designing little windmills, you had to have big ones, otherwise you know a 1,000 megawatt station, even if it was a 2 megawatt design, would require 700–800 windmills. (Interview with David Lindley, 19 March 2009)

There are engineering problems with trying to go directly from small machines to very large ones. Large machines put much heavier stresses on materials than small machines, so it is impossible simply to 'scale up' unless the stress problems are solved (interviews with Donald Swift-Hook, 26 March 2009 and David Lindley, 19 March 2009).

Indeed, by the end of the 1980s there was growing support for renewables, and when the electricity industry was privatised in 1990 the Government included some incentives for a small renewables programme. This was the renewable section of the 'Non-Fossil Fuel Obligation' (NFFO). The NFFO was established initially as a nuclear obligation to make the newly privatised electricity suppliers buy power from the nuclear generators. There was a competition to decide which renewable energy schemes would receive contracts, the results being decided by who bid the cheapest prices to supply different types of renewable energy.

There was a lot of pressure from the public to take renewables forward, putting renewables into the NFFO was not a great cost to anybody. At that time the nuclear obligation was 10% of electricity prices and renewables was asking basically for 1% so there was no great downside to including it…. (interview with leading civil servant dealing with energy policy, 12 November 1999)

The policy was not a major government programme, and the issues of contracts were intermittent, with the Conservative Government showing reluctance to overcome objections from the landscape protection interests, who were critical of the emphasis on wind power. By the time the Conservatives left office in 1997, the strongest renewable technology in terms of capacity was landfill gas, which grew to provide around 1 per cent of UK electricity. Nevertheless, by the middle of the 1990s interest in renewable energy had spread from small companies to the major electricity companies. They established, or bought, renewable development companies. Meanwhile, pressure built up within the opposition Labour party for a substantial commitment to develop renewable energy. Initially a target of 10 per cent of renewable energy by the year 2010 was suggested (DTI 1999a, 34). This came into operation in 2002 and involved an obligation on electricity suppliers to supply increasing proportions of electricity. The size of the obligation was increased after public campaigning by the renewable trade associations, supported by environmental groups. The obligation was raised to 20 per cent by 2020, and then the Government suggested that it would raise the obligation further (DECC 2009, Chapter 3).

The idea behind the Renewables Obligation was that generators would sell renewable obligation certificates (ROCs) to the suppliers, who would suffer a financial penalty for each MWh of their obligation that they failed to supply at any one time. It was originally billed as a market-based system whereby the cheapest sources would fill the target. In fact, as argued in this book, support for renewable energy is based on identification with specific technologies, and there were complaints that technologies such as offshore wind power and solar pv were being underemphasised by the RO because they were more expensive than onshore wind. Delays in achieving planning consent, and delays in upgrading the grid (also affected by planning opposition), meant that the RO targets were not fulfilled.

The RO was then steadily reformed to act, in practice, more like a feed-in tariff system, with different technologies being awarded different numbers of ROCs. Indeed, in 2008 the Government agreed to establish a system of feed-in tariffs for smaller schemes from a number of technologies, including solar pv. Hence it is since 2002 that the UK renewables programme has begun to advance, expanding from around 2 per cent to around 7 per cent of electricity (including the 1 per cent from old hydro) by 2010. In 2007 the Government announced a very ambitious offshore wind power programme that would, if fulfilled, lead to around 20 per cent of UK electricity being supplied from this source alone by 2020. This offshore wind timetable seems unlikely to be entirely achieved, but the target of 20 per cent of electricity from renewable energy by 2020 seems achievable. Indeed, the renewable energy programme has widespread, consensus, support that extended from the Labour Government to the coalition government that took office in May 2010. In 2009 onshore wind was still providing the biggest share of the renewables programme.

Discourse and interests

When the first 'NFFO' renewable programme was launched in the UK there was little political possibility for a Danish-style feed-in tariff. This policy, first adopted by the Thatcher Conservative Government, carried on under New Labour. This coincided with a wider dominant 'neoliberal' free market discourse. As a civil servant responsible for renewable energy policy commented:

> To conservative ministers there was really no significant possibility of there being a fixed tariff for a number of reasons. One was that they were rather suspicious about whether renewables could be competitive and they believed in competition; if you wanted to get the price down then competition was the way to go ... It's just been carried through by the current government. (Interview with leading civil servant dealing with energy policy, 12 November 1999)

The feed-in tariff option was not included as an option by the Labour Government in its consultations about the establishment of the Renewables Obligation in 1999–2000. The only options considered were emissions trading or taxes, direct grants and an obligation, with the last receiving by far the most attention (DTI 1999a, 55–66). Feed-in tariffs were not considered:

> because it did not fit with the market competition ethos of the government. In most REFIT schemes the prices that were to be received were defined by some part of the government mechanism. That was one of the aspects of a support scheme that the British government was anxious to avoid. They wanted the prices to be set by the market and not decided by the government machine itself. (Interview with leading civil servant dealing with energy policy, 30 January 2008)

Some may argue that the obligation route was considered because it favoured the interests of the big companies. Certainly, certificate-based systems do favour big suppliers, who can cope with the uncertainty about future certificate prices much better than small independent companies who need secure power purchase agreements in order to raise finance from banks. However, at the time there was little evidence of any clear preference by the major electricity suppliers about particular options; indeed, two major suppliers were against a supplier-based obligation (DTI 1999b, 8–9). Moreover, there was much less pressure from the sort of constituency that favours feed-in tariff schemes, that is, community schemes and other independent developers, than there is in Denmark and Germany. As a civil servant put it:

> The reason they have local schemes in Denmark are multi-various. One, while Scandinavia is a much more community oriented place,

they are used to doing things in local cooperatives, that's their state of mind, that's their structure and culture, they're used to doing things that look at cooperatives. And they've had local cooperatives for local energy supply continuously. Whereas in the UK we come from a background of nationalised industry supplying energy. So we come from opposite ends of the pole on this issue. (Interview with leading civil servant dealing with energy policy, 12 November 1999)

The campaign for feed-in tariffs only gained a fair head of steam behind it when the success of the German programme of supporting solar pv was heeded and taken up by campaigners rather later, in 2007 and 2008. Rather, the demand from the renewables industry was for some sort of differentiated support, for example through banding. This was a common demand among the various renewable energy interests, and this demand distinguished the position of the renewable energy associations from other interests (interview with former BWEA official, 3 October 2007). The government ignored the entreaties in favour of banding that came from the renewable energy lobbyists when the Renewables Obligation was established:

We believe that a banded Obligation would segment the market unnecessarily, and would lead to Government dictating the relative importance of each technology. We also feel that it is no longer Government's job to pick winners or to introduce artificial distortions into the marketplace. (DTI 2000, 3)

This was an ideological discourse which in fact conflicted with the expressed wishes of the renewables lobbyists themselves. The Government policy was only amended, in 2008, to introduce a series of bands after political criticism that the scheme favoured onshore wind, but not offshore wind and other renewable technologies.

Reforms to the RO to make it function more effectively were discussed at the same time as energy rose very rapidly up the UK political agenda. By 2008 the dominant energy discourse was one of energy crisis, which was widely constructed as one involving growing energy dependence and dangerous levels of energy insecurity. Concern about energy among the British public increased more rapidly than in practically any EU state. According to the 'Eurobarometer' survey, in 2008 19 per cent of the British public regarded energy as one of the two most important policy issues. This was a higher score than any other state in the EU apart from Malta (Eurobarometer 2007–10). The background to this was that production of gas from the North Sea had peaked and was in decline, and net imports of natural gas were rapidly increasing. Natural gas had become the biggest single fuel source in the UK, and at the same time the UK's liberalised gas market arrangements exposed British energy consumers to the full force of energy price increases around this period.

The perception of energy crisis was associated with a avowed move towards a more interventionist policy in government. Indeed, the Department of Energy and Climate Change was brought into being, in 2008, some 16 years after the last Department of Energy had been abolished under the notion that privatisation and liberalisation would lead to market-led decision-making. The discourse about intervention to secure British energy supplies linked up with the discourse about the need to counter climate change to act as a powerful resource for renewable energy interest groups. The Energy Minister responsible for renewables in 2008–10 said, of their relationship with the energy regulators, the Office of Gas and Electricity Management (OFGEM):

> Both Ed (Miliband, the Secretary of State) and I were much more interventionist in terms of approach. We took the view that in order to ensure that we were able to make an energy revolution, that we had to get OFGEM to stop being pedantically market driven. (Interview with Mike O'Brien, 23 June 2010)

Identity EM

The coverage of the early policy on wave power and R&D programmes illustrates how the British renewables programme was stymied for many years because of its reliance on the mainstream energy industry, which did little to develop the industry commercially. Wave power may be thought to be much more relevant to the UK than to other countries, but in fact Denmark did suggest an alternative approach was possible that engaged actors outside the dominant electricity players.

> The Danish support for new wave projects is in extraordinary contrast to all the others... Anybody with an idea for wave energy conversion is given the Kroner equivalent of £5,000. It does not matter if the idea seems bad. It does not matter if it has already been tried and failed several times before. It does not even matter if the inventor is Danish or not. They have set up a small-scale test site at a fiord at Nissum Bredning in North Jutland with a pier, power-supplies,

data-logging, analysis software, shelter, hand-tools, moorings and wave gauges. Anyone who turns up gets the money and can, literally, dive in and get to work. Inventors of devices which show the slightest degree of promise are then given larger amounts of money for tests in the controlled conditions of university test tank facilities. (Salter 2000, 2)

This Danish approach was discontinued after the victory of the right-wing coalition in the election of November 2001, but it does illustrate how a strategy that is less reliant on the major industrial companies, and which encourages independent players, may be plausible.

The 'mainstream' approach to EM might see environmental objectives, such as reduction of air pollution, as coming from the public and the technological means of achieving this left to the energy industry. Although the UK renewables programme has been mainly implemented by companies that are owned by the major energy utilities, it is a programme that has emerged and been sustained through public identification with the worthiness of a renewables programme, and also of particular technologies. The public is most keen on some of these, such as solar power, which may explain how it is that giving high tariffs for solar pv is tolerated politically.

Coalitions between the main trade associations and NGOs have been organised independently of the interests of the electricity industry and, on some key issues, in opposition to the interests of major electricity suppliers. According to a former official of the British Wind Energy Association (BWEA): 'we would use Greenpeace to say messages that we couldn't, so that we could maintain our "let's not rock the boat"' (interview with former BWEA official, 3 October 2007).

One example occurred in the autumn of 2005, when the Government published an Energy Review favouring nuclear power, and in October a junior Government Minister, Lord Sainsbury, suggested that nuclear power could take a share of the Renewables Obligation. 'Suddenly from...the energy review favouring nuclear the potential was there for a nuclear obligation and that nuclear that would take renewables' share of the market' (interview with former BWEA official, 3 October 2007). Greenpeace then took direct action, which disrupted a presentation being made by Tony Blair, then the Prime Minister.

The element of identity EM concerned with idealistic action to develop renewables has, until recently, rarely been seen in action in the UK. At the beginning of the UK renewables programme there were few idealistically inspired initiatives to develop grass roots schemes or start industrial companies. Some activists with a background in the cause-oriented Centre For Alternative Technology did become very much involved in the industry that developed, especially through the company 'Ecogen', but this was mostly concerned with conventional business models rather than community initiatives. There have been a few community wind farm initiatives, but these form a very small proportion of total installed capacity.

However, in 2007–8 a number of 'cause' NGOs were enrolled in a campaign for feed-in tariffs, thus representing a major (successful) effort by popular campaigning to achieve new incentives for small renewable schemes. The successful campaign for feed-in tariffs for small projects (under 5 MW) was organised by the Renewable Energy Association, with Friends of the Earth playing a leading role. This campaign featured support for solar pv as an important element, and the tariff achieved (around 45 eurocents/KWh) is roughly at the same level as comparable programmes on the continent. This was a campaign that won its objectives largely through lobbying MPs. So we can say that 'even' in the case of the UK at least three out of five characteristics of 'identity' EM are achieved. There is the possibility that there will even be a move towards the characteristic of much deployment by independent generators with the implementation of the feed-in tariffs. Indeed, in recent years, with energy security occupying a prominent place in British priorities and incentives for renewable energy, especially offshore wind power, being strengthened and identified as a key way of protecting British energy interests, the British score for meeting 'identity' EM characteristics has increased.

Australia

Birth and development of renewables programme

The development of incentives for renewable energy has been delayed more than in the three previous cases. One factor that stands out is that Australia has, and continues to enjoy, great energy independence. Mining and mineral processing are also an important part of the Australian economy, and industries such as aluminium smelting are intense energy users. This all militates against programmes aiming to cut carbon emissions and raise costs of energy to big industrial players.

The inception of the Australian renewables programme dates back to 2000, when the Mandatory Renewable Energy Target (MRET) was adopted. This is a 'market-based' green electricity trading scheme, operating on much the same principle as the UK Renewable Obligation (as originally formulated so far), but initially it had only a 2 per cent target for the obligation to supply renewable energy. MRET was adopted in the aftermath of pressure to respond to the adoption of the Kyoto Treaty in 1997 (Kelly 2007, 330). Indeed given the considerable degree of energy security (at least in the electricity sector) enjoyed by Australia, this emerges as the chief pressure point for a renewables programme, and the fact that it was more or lessthe only powerful argument compared with much less energy-secure states such as Spain perhaps attests to the relative tardiness of Australia in adopting a renewable energy programme. The 2007 Australian General Election saw a change of Government, with Labour being elected on a pledge to greatly expand renewables. Even the Electricity Supply Association of Australia (ESAA), which is heavily influenced by coal power station interests, shifted from a position of opposition to enlarging MRET in 2003 (ESAA 2003) to one of cautious support for the 20 per cent target in 2007 (ESAA 2007, 2009). However, the ESAA remains opposed to the idea of feed-in tariffs (ESAA 2009, 15).

The proposal to enact an extended version of MRET was passed by the Australian Legislature in August 2009, and it extends the target for supplying renewable electricity to the equivalent of around 20 per cent of total Australian electricity. Various issues remain to be solved. One is that the MRET incentives effectively support mainly one technology, onshore wind, and not technologies including solar pv, whose incentives were more or less cut off in 2008. Even in the case of wind power, issues such as systems for upgrading grid connection between the Australian grid and the other parts of the National Electricity Market need to be solved (interview with Andrew Dickson, Wind Prospect, 5 August 2009).

Discourse and interests

As implied earlier, the discourse that has largely underpinned the moves towards renewable energy development has been focused on combating global warming. The Australian Conservation Foundation (ACF) has taken a lead in campaigning for the adoption, by the Australian Labour Party, of the 20 per cent 'enhanced' MRET. Peter Garrett, a former campaigning pop star who became President of the ACF, became a member of the Labour Government. An ACF spokesperson commented:

> There has been a long and steady campaign to first of all build support for renewable energy policy within the Australian public and there has been lots of polling to suggest that the message has come

through, the Australian public really does identify renewable energy as one of the solutions to climate change. By now, there is the majority of Australians, about 90%, do want to see action on climate change, also one of the solutions that they still have identified, a common thing, renewable energy. That is somewhat a result of lots of different campaign efforts by lots of organisations over many years to build support for that technology. (Interview with Tony Mohr, 29 June 2009)

The ACF have worked with a coalition of environmental groups including Greenpeace, World Wide Fund for Nature and state-based Conservation Councils.

Pressure from the aluminium industry and other intensive energy users in opposition to expansion of MRET has led to those interests being exempted from paying the incentives to support the MRET. Although coal interests, which dominate the electricity generation sector, have now accepted the growth in renewables under the enhanced MRET as not being against their interests, they place a lot of lobbying importance on pressing for carbon capture and storage technology, including considerable expenditure on demonstration projects. There is a considerable difference in priorities in this respect between environmental campaigners and the interests of the electricity industry. 'Basically that (CCS) is part of the future of their industry, as the mining industry for coal, whereas we have a much more I suppose apathetic view about CCS' (interview with Tony Mohr, 29 June 2009).

The amount of emphasis placed on CCS as a solution relative to some renewable interests has also been associated with tensions with other renewable interests. While the major electricity companies have moved into direct promotion of a trade association, the Clean Energy Council (CEC), that promotes renewable energy interests, renewable lobbyists for technologies such as solar pv and geothermal energy have felt that their interests were being sidelined. The CEC therefore split and a new organisation, the Renewable Energy Alliance, has been formed to promote non-wind renewable interests. A Green Party Senator commented:

We used to have a renewable energy association and we used to have a sustainable energy council, then they all merged into this new council and they gave up the word renewable in exchange for clean energy council. That to me signals that they have just been taken over by clean coal and gas. (Interview with Christine Milne, 21 May 2009).

An executive of a solar power company commented:

> There is a consistent conflict between the interests of the small renewable companies, including Solar Power installers and Coal interests as represented now exclusively by the Clean Energy Council. They discuss emissions trading targets and want compensation for the coal industry for the establishment of a carbon emissions trading system ... You have got Dracula in charge of the blood bank, in effect. Solar interests were sidelined. (Unattributable interview with executive of solar trading organisation, 14 August 2009)

A spokesman for Wind Prospect, an affiliate of the CEC, is more supportive of the CEC. The CEC was active in lobbying for changes that were adopted by the Government:

> It is fantastic to have a credible and active peak organisation in the shape of the CEC to represent renewable energy in Australia. Two years ago the CEC was formed out of an amalgamation between the Business Council for Sustainable Energy and the Australian Wind Energy Association (AusWind). A number of members of the CEC have renewable generation assets but also gas powered generation assets. The CEC's approach has been one of private lobbying behind closed doors rather than grandstanding in public and this appears to have produced results in this case. Other environmental groups have adopted a more public approach in their push for renewable energy and green jobs. Both these types of approaches are necessary and complement each other. (Interview with Andrew Dickson, Wind Prospect, 5 August 2009)

The wind developers have supported a certificate-based system as long as they can arrange power purchase agreements with the suppliers, who in turn have become supporters of wind power development through investments. On the other hand, supporters of solar power and other renewables that are more expensive than onshore wind have supported feed-in tariffs. So far little headway has been made in this regard, apart from a scheme in the Canberra area under the aegis of the Australian Capital Territory.

Identity EM

In terms of the five characteristics of 'identity EM', we can see that in Australia there is a mixed result. There has been relatively little activity on an idealistic basis by individuals or groups. Australia is now developing a major renewables programme, but this is being done mainly through the electricity majors in the context of an existing, multinationally based, renewables industry, especially in the case of wind power. Compared with the other case studies, Australia is a late entrant into the large-scale renewables industry. However, some key aspects of identity EM are present, including a dedicated renewable incentive mechanism (MRET) as well as some subsidy schemes supporting solar thermal domestic installations. In addition, the environmental groups have launched campaigns dedicated to the support of renewables rather than the preferred mainstream industrial solution to carbon dioxide reduction, namely carbon capture and storage. They have therefore been in a coalition of interests with the renewables sector.

Other indicators of identity EM are more ambiguous, in that the big trade association representing renewables, the CEC, seems to have much more direct influence from the major energy companies themselves rather than substantial existing renewable development companies, as tends to be the case in other countries with major renewable programmes. In other countries the renewable development companies have their own corporate identities, which are distinct from the organisations that may own them. The Australian relationship between the energy majors and the CEC has led to an alleged clash between at least some renewable energy and the core interests of the major energy generators involving gas and coal.

It may be argued that a 'mainstream' EM approach may have more to say about Australia than cases such as Germany, Spain or even the UK. However, in so far as it does, this may suggest a reason for the relative slowness of the development of renewable energy in Australia. Nevertheless, some key characteristics of 'identity' EM are still present in the Australian example, and it is clearly the case that the commitment that has been secured to promote renewables through MRET is one that has been secured by public campaigning for, and public identification with, renewable energy, rather than a purely independent decision by the main energy companies to dramatically expand the renewables programme.

China

Two key research questions emerge for this case study. First, to what extent are patterns of policymaking on renewables different from those observed in the West? For example, is it correct simply to describe the

Chinese policymaking in renewables as 'authoritarian' compared with the way things are done in the other, Western, cases described? Second, what results are obtained when 'identity' EM analysis is applied to the Chinese case? This is answered in what is an initial, outline study, by mainly referring to the wind power industry.

There have been some analyses of Chinese policy on renewables in English-speaking journals (Lewis 2007; Li et al. 2006; Quiang Wang 2010; Xiliang Zhang 2010), but there has been little analysis of the politics of policymaking in this field, Mah (2010) being one example of political analysis of renewables. Casual analysis may assume that China is not susceptible to the sort of interest group and EM-style analysis with which environmental policy is studied in Western states. Indeed, it is possible to assume from some statements about Chinese policy styles that it is necessary to say just that policies are decided by a central authoritarian communist party with little elaboration needed. One noted political scientist summarises some work on the politics of China by saying:

> Not only has China's expanding bourgeoisie shown little interest in achieving rapid democratisation, but the emergent capitalist class has demonstrated a willingness to work with and support the non-democratic, authoritarian Chinese Communist Party ... Indeed, there is compelling evidence that the state is becoming *more* powerful rather than less as the government seeks to reassert greater control of the economy. (Beeson 2010, 16)

China has recently embarked on a major wind power deployment programme. Indeed, installed capacity jumped in 1 year from 12.5 GWe at the start of 2009 to around 25 GWe in early (Qi Wu 2010). Are we to assume that this is achieved simply by top-down instructions from the leadership, down a bureaucracy to the electricity industry? Such a prognosis would certainly appear to be an antithesis of the sort of 'identity' ecological modernisation that has been discussed in this book – indeed, it would be an antithesis to much of the market-oriented nature of mainstream EM itself. Beeson (2010, 18–19) even appears to argue that environmental authoritarianism is a necessary feature:

> The central question that emerges from this discussion is whether democracy can be sustained in the region (China and South East Asia) – or anywhere else for that matter – given the unprecedented and unforgiving nature of the challenges we collectively face, such is

the urgency of the environmental crisis that ... The persistence, if not the intensification of authoritarian rule in the face of environmental decline and reduced expectations about the course of economic development seems an all too likely outcome as a consequence. (Beeson 2010, 18–19)

A basic assumption underpinning all this analysis is that it is taken as virtually read that environmental policies, such as they are, in China are imposed by authoritarian rule and so have little to do with the sort of liberal practices analysed in the West. Indeed, it is sometimes emotionally difficult to argue with this proposition since, in Western eyes, this almost seems tantamount to being some sort of fellow traveller excusing or glossing over the treatment of dissidents who question the legitimacy of communist rule. It is certain that many accuse the Chinese government of acting in an authoritarian manner in the field of energy, for example in the form of the massive 'Three Gorges' Dam project in the Yangtze River. Two points can be made here. The first is that there has been resistance to the project, and internal questioning by people arguing about the ecological consequences. The second point is that such experiences do not necessarily translate to other policy areas.

In fact, the issues of how policy is decided in discrete policy areas and the issues of democracy and nationality are distinct. It should be understood that, if the Chinese state tolerates open discussion in at least some discrete policy areas (although not questions about the legitimacy of the regime), then it may indeed be possible to talk about interest group politics in a Western sense. There is evidence for this in the field of development of policies for 'new' renewable energy sources such as wind power. Generally, however, energy politics is more complex than a simple authoritarian model would suggest. As a leading energy analyst puts it: 'Most of the world when trying to understand energy politics in China thinks it is central planning and the Chinese government has got it all under control and they can do better than the West. Actually it is quite the opposite' (interview with Sebastian Meyer, 29 October 2009). Indeed, there is growing evidence of local interests taking action to build wind power developments before the grid infrastructure has been organised through central decision-making. In 2010 around a third of the wind power capacity had not yet been connected to the grid (Qi Wu 2010).

In fact, Mol (2006), who has spent some time analysing China, concurs with the view that the Chinese case is becoming more complex. Mol aimed to assess the extent of European-style ecological modernisation in China by investigating 'the growing autonomy, independence or "differentiation" of an ecological perspective and ecological rationality vis-à-vis other perspectives and rationalities' (Mol 2006, 33). He concludes that:

> China's strategy and approach to tackling the growing environmental side-effects of modernization is far from stable; it is still developing and transforming, together with the general transition of China's economy and state. But most environmental reform initiatives are firmly based on, make use of and take place within the context of China's modernization process. In that sense, it seems justified to use the term 'ecological modernization' to describe China's attempts at restructuring its economy along ecological lines. (Mol 2006, 51)

Certainly China is still emerging from a centrally planned economy. Until 2002 energy decisions were taken by the Energy Ministry. While decisions on energy prices are still subject to regulation by the state at various levels, the decision-making structure has changed and the ownership patterns of the industry are now more complex (Rosen and Houser 2007; interview with Sebastian Meyer, 29 October 2009). There is now an increasing amount of private ownership, although, as a leading Western China energy analyst puts it, state ownership is manifest in 'not just the national government, you can clearly say that is state owned, but provincial levels, county levels, city governments, own all kinds of intertwining equity states in businesses' (interview with Sebastian Meyer, 29 October 2009). Since 2002 decisions on renewables (as with many other aspects of economic development) are taken by the National Reform Development Commission (NRDC). Within that, provincial government is also important. According to Rosen and Houser (2007, 25):

> Provincial officials lobby for end-user pricing low enough to keep their industries viable and citizens happy. The power generators lobby for an on-grid tariff high enough to cover their fuel costs and ensure that the sector is profitable enough to make the necessary investments. In the middle are the grid companies, stressing to NDRC that enough margin needs to shake out between the two prices to finance a $130 billion buildout of the nation's transmission network between 2006 and 2010.

In other words, there exists the sort of interest group politics that is immediately recognisable from political science studies of Western states. It is not as if central planning and price-setting are unheard of in the West. In the UK it was not until 1990 that any significant private ownership of the electricity industry was allowed. It was only after privatisation took effect in 1990 that the price-setting was devolved from the state-owned centre to a market. In France, this centralised state-owned condition continued even later, and arguably still exists to a great extent today, with the state maintaining a 35 per cent stake in Electricité de France.

China now has some (relatively) ambitious targets for wind farm development – 170 GW of installed wind capacity by 2020 out of around 1200 GW of projected total electricity-generating capacity (Ngan 2010, 2143–4). This will mean around 5 per cent of electricity from wind power.

Far from wind farm development simply being imposed from above, there appears to be considerable pressure coming from below for wind power development. [The] 'Local level wants wind projects, the wind projects bring a lot of investment' (interview with Sebastian Meyer, 29 October 2009).

Also, according to a Greenpeace spokesperson:

[S]o if the Local Government can introduce more renewable energy projects into the area, that means they can secure the more job opportunities as well as funding from other Central Government and those developers…Say Inner Mongolia, that is the biggest wind power installation (area) in China…All those installations, transportation and maintenance will have job demand (produced) in that area. (Interview with Liu Shuang, Greenpeace China, 23 December 2009)

Indeed, it was in the provinces that feed-in tariffs first emerged, for example in Guandong province (Mah 2010), before the central government started a system of competitive bidding for projects, which has now given way to the feed-in tariff system. Shandong Province is giving feed-in rates that are above the rates set by the central government (interview with Sebastian Meyer, 29 October 2009).

The requirement that 70 per cent of procurement value of wind power installation go to local companies has proved controversial in the West, and has been condemned as 'protectionism' by Western wind generator manufacturers and their political supporters (Wu Qi 2009). Yet Chinese renewable energy advocates (including Greenpeace) believe that maximising local job opportunities is a good idea. In fact, this Chinese policy of preference for local manufacturing is strikingly similar to the policy in Spain. Just as Chinese provincial governments have pushed for wind development, so have the Spanish regions. Indeed, Lewis (2007, 18) has recognised the similarity between Spain and China in this respect. One could also add that in both countries landscape objections are rarely an issue of objection to wind farms. Both countries have a relatively poor rural economy that leads to a premium being placed on schemes that involve economic development. How does the China case fare when seen through the lens of 'identity' EM analysis?

Identity EM

Idealism

There is little in the way of the innovatory idealism that marked the wind power technology and implementation movement in Denmark, Spain and Germany in the 1970s and 1980s. However, the same could also be said about the cases of the UK and Australia discussed in this chapter – not to mention the many industrialised, as well as developing, countries with even less in the way of renewable energy programmes.

Financial support mechanism

Since July 2009 China boasts a feed-in tariff, one that is varied according to the wind regime in different regions. The feed-in tariff for wind is at least 50 per cent over the prices given to coal-fired plant. In this sense an important criterion for 'identity EM' is in place, in that there is a dedicated premium price mechanism for wind power. In that sense it has a characteristic in common with the other cases covered in this chapter.

Independent trade associations

The Chinese Renewable Energy Industry Association (CREIA) was established in 2002. It represents a range of manufacturers and developers of different renewable fuels. These include wind power but also biomass, biogas, solar pv, solar thermal, small hydro and also ocean energy (Li et al. 2006). CREIA is independent of the main electricity companies, something that is not clearly the case in Australia with the Clean Energy Council.

Coalitions with NGOs

CREIA has been active in forming a coalition with NGOs such as Greenpeace and WWF in calling for more renewable development and for feed-in tariff systems (Li et al. 2006). Greenpeace's pro-renewable campaign has involved some hard-hitting attacks on the impact of the coal industry in China (Mao Yushi 2008). In order to push the campaign for feed-in tariffs, Greenpeace organised a series of seminars involving representatives of the renewables industry and leading policymakers and bureaucrats. These included personnel from the Energy Research Institute (ERI). The ERI has close links with the NRDC, which puts it in a key policy position. Greenpeace can boast influence to the extent that it has produced policy documents in collaboration with personnel from the ERI (interview with Liu Shuang, Greenpeace China, 23 December 2009).

Independent developers

There is little independent development outside the mainstream electricity industry, although there are the beginnings of project-financed schemes in a few cases (interview with Sebastian Meyer, 29 October 2009). However, one thing that China does have in abundance in terms of independence is a very strong and rapidly expanding wind generator (and solar pv) manufacturing industry. Indeed, in the relatively near future the Chinese wind power manufacturing industry is likely to be challenging the European and US generator manufacturers for market share in the Western markets themselves.

The answer to the question about whether renewable politics are comparable to what are often regarded as being 'advanced' Western cases is that they are comparable. The use of the identity EM analysis certainly seems to point in that direction. Indeed, as will be discussed in the conclusion, not only does renewable politics appear to have a more independent and effective lobby than in the case of Australia (and, arguably, also in the UK), but it also may involve greater 'bottom-up' pressure for development.

Conclusion

In the five cases studied, major elements of indicators of 'identity' EM (as discussed in Chapter 2) can be observed in all five cases. However, the extent of identity EM varies according to the cases, with Germany having the clearest examples, while Australia (note, not China) has the least. Germany and Spain, which both saw the wind industries formed largely on the back of often idealistically inspired initiatives, are the countries that provide the most extensive programmes of renewable energy to date. In Germany the renewable energy industry remains mainly independent of the main electricity industry. This is not the case in Spain, although in this case the renewable industry retains and expresses different interests and identities compared with those of the mainstream energy industries.

The UK fits in with at least three out of five of the characteristics of identity EM, whereas the case of Australia is more complex. The biggest trade association seems not to have the same degree of independence from the established utilities or to engage in public debate compared with the trade associations in other cases. Curiously (if we reflect on the stereotypes), China seems to have a more independent (of the energy utilities) renewable trade association and to have rather more effective coalitions between the renewables sector and the environmental NGOs than Australia. Not only does China have a renewable trade association that is more independent than that existing in Australia (and, indeed, even in the UK), but also there is bottom-up pressure for renewable developments. This bottom-up pressure appears to be greater than in either Australia or the UK, where there is little of an indigenous wind turbine manufacturing industry. Of the case studies under consideration, Australia seems, according to the framework considered in this book, to have most association with mainstream EM. However, even in this case, the existence of a dedicated incentive mechanism for renewables (MRET) and the coalitions between environmental NGOs and renewable trade interests demonstrate that there is clear evidence of (some) identity EM characteristics.

However, we must be careful not to lump all renewable energy technologies under stereotypical conclusions even for the same country. Solar pv exerts an identity that remains independent of the conventional electricity suppliers, even in the case of Australia. This technology is represented by different renewable associations from those of wind in all of the cases. However, the fact that solar pv relies particularly heavily on the politics of ecological identity may also be associated with the fact that it is at a relatively early stage of development compared with wind power, which at some levels is beginning to become accepted into the mainstream electricity regime itself. What seems clear overall is that the rapid emergence of renewable industries is associated with 'identity' EM, and slowness in developing renewable programmes seems to be associated with a more mainstream EM approach.

As a conclusion, it does seem possible to tie together the extent of identity EM with the speed and extent of renewables programmes in these cases. The number of 'identity EM' characteristics that a country-case study shares seems related to the size and speed of development of its renewable energy programme. Germany and Spain have the most identity EM characteristics, followed by the UK, followed by China, then Australia. The proportions of electricity generated in these countries from new renewable energy sources are also in that order. In addition, the deployment of less well-developed technologies, especially solar pv, seems also to be in proportion to the extent of the relevance of 'identity EM' characteristics. The case with most evidence of mainstream EM, Australia, has been the slowest to develop. Mainstream, as opposed to 'identity', EM is almost a predictor of lacklustre renewable energy deployment.

However, if we are to try to explain the rate and extent of renewable development in each of the cases, we also have to take into account other dimensions, such as dominant discourses and perceived interests. In Spain, for example, it is not necessary, as in Germany, for there to be evidence of the 'identity' EM characteristic of much renewable deployment by independent companies. This is because the dominant energy discourse in Spain involves notions of seeking to counteract dependence on imported fuel supplies. The energy majors thus see their interests as being engaged with pursuing this objective through support for renewable energy, especially since influential discourses allied with institutional arrangements serve to effectively rule out the nuclear power option in Spain. By contrast, in Australia, the dominant discourse has been one not only of successful energy independence through use of coal, but of exporting coal as a major component of national well-being. It is only in recent years, with the emergence of discourses about the threat of global warming, that the energy majors have taken a more positive attitude towards renewables, mainly large-scale wind power. Germany has had a mixture of fears of energy dependence as well as influence from its own domestic mining interests. However, the influence of environmentalists and anti-nuclear campaigners has been very strong. The UK is a case where, in recent years, increasing fears of energy dependence on imported natural gas have helped to drive perceptions of the need for renewable energy up the political agenda. In China there is a discourse among organisers of the electricity industry that bringing renewables, especially wind power online, is necessary to appear as being in the forefront of developmental programmes (interview with Sebastian Meyer, 29 October 2009).

However, there are big differences in the relative contribution that new renewable energy makes to the task of developing sustainable energy supplies. In Germany, in the period 2004 to 2008, the share of electricity supplied by renewable energy increased from 9.7 per cent to 15.1 per cent, an addition of 5.4 per cent to German electricity supplies.

This outpaced overall growth in Germany's electricity of around 3 per cent during the 2004 to 2008 period (Business Insights 2010). In the UK the increase in renewable energy, at around 1 per cent per year, was roughly in line with the increase in total electricity generation, and also coincidental with a decline in carbon emissions from electricity generation (but also an increase in imports of natural gas). Although in Spain renewable energy is increasing rapidly, such has been the growth in Spanish energy consumption, of around 50 per cent between 1990 and 2004, that the proportion of primary energy (and electricity) coming from renewable energy has remained much the same (Bechberger 2007, 217). Australia's historical growth in electricity demand seems likely to be larger than the projected increase in renewable energy to meet the targets set by the Government. In the 1990 to 2006 period electricity demand increased by 65 per cent (World Nuclear Association 2010). In the case of China the position is especially acute. Since 1990 electricity consumption has increased by over a factor of three. Even though this growth may slow and Government targets for renewable energy may sound big, they will not come close to keeping up with increases in electricity demand.

Identity EM may be able to boast modest successes in Germany, in absolutely decreasing the need for electricity from non-renewable sources, and (more arguably) in the UK, but not in the other cases studied here. As will be mentioned in the concluding Chapter 7, if identity EM is to work as a strategy to deal with sustainable climate change targets, then it needs to be both strengthened in the renewable energy sector and applied to other energy sectors.

7
Conclusion

Using 'identity' EM

I shall begin with some concluding comments about ecological modernisation (EM) theory, since that is the key theoretical focus of the book. Later on I shall make some comments about the use of institutionalist theory. These theories (EM and institutionalism) are linked in this case because so much discussion is at the level of what policy instruments, and what policy institutions and incentive structures, are required, when in fact the subtext of this discussion is about technological choices, with different interest groups vying for policy instruments that favour them. It is notable that big energy corporations often stress the importance of 'market-based' trading mechanisms, whether it be emissions trading or renewable certificate trading systems. This fits in with what I call the 'mainstream' EM approach, which involves the mainstream industries taking on board environmental concerns about pollution reduction but reserving to themselves the decisions about which technological options should be deployed to achieve the environmental objectives. Not only can the incumbent industries use the incentives to ensure that most of the technology is owned by them, but they can choose types of technology, for example nuclear power, or possibly carbon capture and storage (CCS), that might not, especially in the bulk of the EU, be a priority choice from general publics for subsidy provision.

An 'identity' EM approach, as discussed in this volume, is one that is more associated (than mainstream EM) with accommodating popular technological preferences. The central political relationship here that links the alternative (renewable) industry to environmental NGOs and movements is the popular identification with the alternative technology, in fact specific technologies such as wind power or solar power. Hence this book talks about the notion of 'identity ecological modernisation'. A contrast between 'identity' and 'mainstream' EM may be typified by the debate between protagonists of 'feed-in tariffs' and emissions allowance or green electricity certificate trading systems, devices favoured by transnational electricity companies who claim to make the best, most rational, cheapest choices on our behalf.

The argument set out in the preceding chapters is that, at least in the formative and developmental stages of the renewable energy industries (that is, up until now), there has been a rough correspondence of the size of national renewables programmes with the extent to which we can observe characteristics of 'identity EM' in these countries. It is no surprise that the largest renewable energy programmes tend to be associated with feed-in tariffs. The USA appears superficially to be an exception to this rule, but, as is discussed in Chapter 5, in fact the USA is not an exception because the incentives for renewables in the USA rely mainly on state-based traditional regulatory measures and bargaining. These are supplementary to the Federal Production Tax Credit, which acts in the same way as a small feed-in tariff to favour wind power in windy places.

As has been discussed in Chapter 2, the identity EM approach discussed here is to be distinguished not only from mainstream EM but also from the sort of 'deliberative' EM put forward by Hajer (1995). In criticising the latter, Hajer-esque formulation of EM the intention is not, incidentally, to distinguish the approach in this book from the poststructuralist methods employed by Hajer. On the contrary, a problem with Hajer's analysis is that there may be an implicit belief in some sort of universalist ecological consensus that may be uncovered by a sufficiently rigorous deliberative process. Deliberation is a value that is a 'good' in itself in the value structure of environmental justice, but support for it should not blind analysts to the existence of different identities with ecological and technological values. As has been discussed in Chapter 3, we can often understand the planning politics of renewable energy, and analyse the different outcomes in different places and times, by talking about different, sometimes conflicting, identities that influence the outcome of debates about renewable energy. Ironically, Hajer and others who talk about discursive deliberation seem to be implicitly searching for some consensual, universalistic 'truth' of discursive difference rather than recognising a clash of identities and traditions as a basis for analysis and normative problem-solving. The talk about 'reflexive modernisation' seems to focus mostly on opposition to technology, and not enough on public support for what many people see as being clean, renewable technologies.

A central aim of this study has been to devise and then deploy the notion of 'identity EM' in order to analyse different case studies and help understand differences in outcomes. Five characteristics of identity EM were identified: idealism; dedicated financial support mechanisms for renewable energy; independent trade associations representing renewable energy; coalitions between renewable trade groups and environmentalists (often in opposition to the position of the mainstream energy industries); and deployment by companies that are independent of the major energy corporations. The countries (studied in this book) with the most evidence of these characteristics include Denmark, Germany and Spain. The country with the fewest characteristics is Australia, with the UK and China forming a middle group behind Denmark, Germany and Spain, but ahead of Australia. The heavily federalised nature of renewable energy policy in the USA makes it difficult to make a generalisation about the USA and identity EM, although analyses of some individual states with stronger renewables programmes have been made. California in the late 1970s and early 1980s seemed to satisfy all of the five 'identity EM' criteria, for example. The other states that have been analysed (Texas, Iowa, Minnesota and contemporary California) appear, to varying extents, to satisfy four of these five 'identity EM' criteria.

The countries with the (proportionate to their country's size) biggest early renewable energy programmes have been associated with considerable evidence of all of the identity EM characteristics. Denmark (up to the end of 2001), Germany and Spain have most evidence of identity EM characteristics, including 'idealism' in the early stages, which involved activists sharing ideas, forming cooperatives with political as well as commercial objectives, and investments being made without any certainty of short-term profit. In California of the 1970s, idealistic campaigners for 'soft energy paths' were influential through the establishment of the Office for Appropriate Technology. Idealism is more typical of earlier, pioneering programmes, before the period when the major electricity companies decided to invest heavily in renewable energy.

In Spain, and to a growing extent in China, the mainstream energy industries are supportive of renewables to a greater extent than in Germany or in California, where it is independent companies, supported by environmentalist pressure, that have mainly developed the renewable industry. This contrast is associated with greater feelings of growing energy shortages and also, in the cases of Spain and China, rapid expansion of electricity demand. This means that existing energy companies do not feel that their core interests are threatened by development of renewable energy. Renewable energy programmes that have been slower to develop, such as in Australia, have had less evidence of identity EM characteristics than other countries studied. However, while China does not have such strong identity EM characteristics as Germany or Spain, it has more than Australia, by virtue of having a renewable trade lobby that is perceived as being independent of fossil fuel interests, something that is not always apparent in the case of Australia. Coalitions of environmental groups and renewable trade interests have fought for the establishment of a significant dedicated financial support mechanism for renewable energy in Australia, but this has only been in place since 2009. The strong coal interest has been influential.

Of course, the case studies covered all have, or had, to a greater or lesser extent, evidence of renewable energy programmes. There are still countries that would not fulfil any of the criteria for identity EM, and none of those cases studied here had any of the criteria before the 1973 oil crisis.

Overall, it can be said that identity EM can help us understand renewable energy politics and policy, and assist comparisons between different countries. The identity EM characteristics can be used to understand the success, extent and nature of the renewables programme.

Identity EM and 'rationalism'

Understanding how technological choices are made and how these are linked to policy contexts and debates about appropriate policy instruments is an interpretive exercise, albeit one in which standard 'heuristic' devices (such as the 'identity EM' classification that has been used) can be deployed to categorise the evidence that we need to interpret. Here I attempt to separate out analysis from overtly normative prescriptions. However, the notion of 'identity EM' is explained by somebody (i.e. myself) who is sympathetic to the cause of renewable energy, so in this sense normative and analytical analysis cannot be totally separated.

Many in the energy industry would complain that, however accurate I may (or may not) be in categorising identity EM characteristics of renewable energy programmes, the very idea of letting policy be too influenced by ideas that appear fashionable (that is, popular) with ordinary citizens may allow policy to become divorced from responsible 'rational' consideration about cost and technical feasibility. There are two responses to this. The first that, even if this is accepted as the case, these self-declared 'rationalists' would do well to understand through analyses such as these how the policy outcomes are occurring, or else their chances of achieving their own 'rational' objectives and practices will be much reduced. The second response, and one that I shall spend some time discussing, is that the so-called rational approaches themselves are vulnerable to the criticism that they hide their own reliance on value-laden, non-rational assumptions. Their alleged rationality seems often to rest on the dominant habits and modes of operation of the prevailing technological regime. Of course, as the political and social 'landscape' changes, such rational habits become challenged and supplanted by new technologies (Geels 2002). This appeal to rationality in a way that fits in with establishment industrial interests falls prey to the classic criticisms of scientific rationalism deployed by Habermas (1971). This involves the idea that entrenched interests achieve their political objectives by a challengeable claim to have a monopoly on reason and technical practicality.

As was noted in Chapter 3, renewable energy objectives have been taken on board by transnational energy corporations themselves to such an extent that large commitments to renewable energy deployment are now accepted (though begrudgingly at times). Hence, today, when people represent the electricity industry's preferences they tend to argue from the basis that 'renewables can't solve everything' as opposed to the argument they used to deploy in the 1970s and 1980s that renewables 'can't solve anything (much)'. This does beg the question: that is, if so much that was rational 30 years ago is no longer so clever, then why should we believe the revised 'rationalism' now?

In entering this discussion I move from a path that I have tried to tread as an analyst who is friendly to renewable energy objectives to a more overt position as a normative advocate of renewable energy. This shift in position does, however, help analysis as well, for in debating a normative position it is possible to demonstrate that there is no such thing as a claim to rationality, at least, that is, outside a carefully delineated framework or discourse. Of course, we share a common discourse, which allows us to debate, but this debate is, and should also be explicitly, about values. This ought to be acknowledged, and not obscured behind claims to have sole ability to understand reality.

A critique of (nuclear) energy 'rationalism'

How important is renewable energy? The thrust of a normative, identity EM approach to renewable energy is that renewable energy is too important, too central to energy futures to be left solely to mainstream industry. The technological choices need to be made through public debate and clear incentive structures. Often this happens after political campaigns fought by coalitions of renewable trade interests and environmentalists against indifference or hostility from the conventional energy establishment. Yet this 'identity EM' faces a counterattack from supporters of maintaining a strong role for nuclear power in our energy futures. These advocates, such as MacKay (2009), while increasingly recognising a central role for renewables, claim to represent reason above the emotion displayed by Greenpeace supporters. MacKay states that:

> Twaddle emissions are high at the moment because people get emotional (for example about wind farms or nuclear power) and no-one talks about numbers. Or if they do mention numbers, they select them to sound big, to make an impression, and to score point is arguments, rather to aid thoughtful discussion ... This is a straight-talking book about the numbers. (MacKay 2009, viii)

MacKay's claims to master the high ground of number-crunching rationality can be challenged as being an exercise in itself in number selection. The values or 'emotions' that he claims to set aside are implicit in his own analysis. Indeed, the apparent thrust of his book, that we cannot rely completely on renewables (and that nuclear power is crucial), seems undermined by a plausible interpretation of his own numbers. For example, on page 103 (Figure 18.1) MacKay summarises his findings that a combination of different renewables (largely solar power and wind power) would provide almost all of current UK energy demand – which in itself seems to speak rather positively about UK renewable energy potential. However, it can also be argued that MacKay's presentation of the numbers is a selective one that substantially understates the renewable energy case. For a start, he seems to use a rather high figure for energy demand when comparing it with the potential for renewables.

MacKay uses an unconventional way to count energy consumption, using 'kWh per person per day' as opposed to the more familiar (to energy statisticians) account of TWh or MTOE per year for the whole country (UK). If we compare the actual 2008 UK energy consumption (DECC 2010) of 224 MTOE, converted into 2605 TWh, with MacKay's figure of 195 kWh per day, which comes to 4270 TWh, we can see that MacKay's figure is a lot larger than actual UK energy consumption.

MacKay's justification for this is convoluted. In fact, MacKay does not claim to reproduce UK energy consumption, but talks about an average Western energy consumer. But then he compares this with UK renewable energy resources. Clearly he is not comparing like with like. If he did, then his estimate of total UK renewable energy potential (around 4000 TWh) would appear to be rather larger than UK energy consumption (2605 TWh in 2008). MacKay may argue that we import a lot of embedded energy in our goods, but this is an argument for these overseas producers to be sustainable, not for us to generate their energy for them.

Crucially, MacKay assumes some unreasonably low numbers for renewable energy uptake in practice. According to MacKay, anything more, for example, than 10 GW of offshore wind power would be ruled out because of either radar objections or objections on grounds of bird protection (page 109). This is interesting given that, at the time of writing, something of the order of 40 GW of offshore wind capacity has already been earmarked for development in types of projects that do not usually experience planning objections from radar or the Royal Society for the Protection of Birds. Even including onshore wind, it seems that, according to MacKay, only 20 GW of wind power is plausible (page 214).

MacKay describes the notion of providing large amounts of UK energy through solar farming as 'fantasy time' (page 41) – an emotional response, surely. The 'rational' energy establishment dismissed the early wind power pioneers of the 1970s using much the same language in which solar pv is dismissed today. Of course, much depends on the extent that solar power declines in cost. There is great uncertainty about this, but making an assumption that it will *not* decline greatly in price over the coming decades is to make an 'emotional' judgement just as much as believing that it *will* decline much in price by 2050. Indeed, there are very plausible arguments to doubt the value of offering, as is being done in the USA, incentives for nuclear power rather than using the money to support solar power, wind power or other renewables. Nuclear power is a mature technology, whose costs are unlikely to be lowered, while solar pv is a developing technology, whose industrial maturation will benefit from such financial incentives.

It is a matter of belief, and MacKay seems to believe in nuclear power futures, something that he does not criticise very much. However, in what way, for example, is an apparently uncritical review of the possibilities for provision of major proportions of the world's energy supply through fast breeder reactors justified, given the question marks surrounding the technology (163–6)? Fast breeder technology has, arguably, never been made to work properly, it involves systems that are criticised as being potentially much more dangerous compared (even) with conventional nuclear power, and it relies on a nuclear fuel recycling process that not only gives rise to large quantities of nuclear waste but involves the production of large quantities of plutonium, itself an especially dangerous commodity. MacKay argues that 'if we wanted to crank up nuclear power 40-fold worldwide, in order to get off fossil fuels and to allow standards of living to rise, we might worry that once-through reactors are not a sustainable technology' (MacKay 2009, 163).

If fast breeder technology is itself to be avoided, where does this leave a world increasingly reliant on nuclear power? The answer is that the world will be heading towards 'uranium' supply crises to replace oil crises in decades to come. Is it not incumbent on countries, including the UK and the USA, who could plausibly derive most or all of their energy from renewable sources to vigorously pursue this renewable energy solution rather than build nuclear power stations? They may need to do so in order to ensure that 'uranium' crises do not occur and, it follows, to ensure that the adoption of what many see as the nightmarish fast breeder energy economy does not become a reality. MacKay's nuclear-oriented rationalism is no more immune to the 'twaddle' of emotion than any other form of energy rationality.

Indeed, if the UK, with its densely populated nature, is potentially able to source all or most of its energy from renewable sources, then this is even more likely to be the case with other, less densely populated countries. This includes the USA. A study undertaken by the National Renewable Energy Laboratory for the US Department of Energy found that the 'lower 48' US states have the potential to produce nearly 10 times the total current US consumption of electricity from just wind power on its own (NREL 2010).

One of the implicit aspects of Mackay's model, which is widely accepted, is that electricity is likely to greatly increase as a proportion of final, delivered, energy supply. Just as in the case of the 1970s oil crises, the crises were not fully resolved until energy economies were restructured by shifting electricity production and much industrial and domestic fuel use away from oil. The problem more recently is that demand from the remaining stalwart oil demand sector, transport, has at times appeared to expand more rapidly than the increase in oil supply streams. With the rising appetite of developing countries, led by

China, for new resources, and oil in particular, the world appears to be in the grip of a commodity/resource crisis.

It does seem that, given technological improvements in batteries, it is now more plausible to imagine a structural shift towards significant supply of fuel for motor vehicles through electric cars. Traditionally, in countries such as France, nuclear power advocates have thought that this favours nuclear power, although in much of the world this may, in the near term, this could do more to favour the use of natural gas to power combined cycle gas turbine power stations. However, greater use of electric cars could favour renewables, in a technical sense, more strongly than nuclear power, since electric cars provide a potential for energy storage, and thus help balance demand to cope with the variability of renewable energy sources such as wind power and solar power. It may be that the public will identify electric cars much more with a renewable future than a nuclear future. This is relatively simple to organise by offering tariffs to people to charge their car batteries based on the need of the electricity system for power. When there is a relative shortage of renewable energy, tariffs for battery recharging would be high; when there is a relative surfeit of renewable energy, electricity tariffs would be low.

One important aspect of 'identity' EM, which acts to distinguish it from mainstream EM, is the notion that the public has a view on the technological means of achieving environmental objectives, and not merely the environmental objectives themselves. This aspect has already been discussed in Chapter 2. However, its significance needs to be flagged when pointing out to those, like MacKay, who put themselves forward as rationalists. Consumers consciously choose specific technologies through the political process (and sometimes on a direct, individual level). This leads to public acceptance of incentive schemes that are paid for out of electricity consumers' bills (through feed-in tariffs in particular), or, for instance, in the case of the US production tax credit, are paid out of notionally forgone tax receipts. The relationship between economic development and ecology is therefore subtly different in the case of identity EM compared with mainstream EM. It is still assumed that improvements in environmental quality go hand in hand with economic development, but in the case of identity EM priority is given to sustainable technology, with which the public identifies through positive opinions and active lobbying groups. Economic development is made conditional on its being achieved in the context of the development of these specific ecological technologies.

How institutions are important

In this work institutionalism has mainly been conceived as a means of unpicking the arguments and influences concerning choice of policy instruments to promote renewable energy. The discussion about whether policy instruments supporting clean energy projects should be 'market-based' or 'command and control' exists only to appeal to prevailing political discourses about the supremacy of so-called 'market'-oriented policy. As explained in Chapter 5, Hodgson (1998) makes it clear that markets should be analysed as institutions. When we start doing that, we can better understand how things work (or do not work) for the stated policy objectives. This is as opposed to making the erroneous distinction between so-called 'market-based' and 'command and control' mechanisms. All involve state intervention of one sort or another; the question is what is most effective in maximising the deployment of a preferred technology. A key part of the answer is to make technology choices, and then institute institutional arrangements that give the necessary confidence to investors and lenders of long-term returns to their capital investment. Feed-in tariffs have often proved to be an effective means of promoting renewable energy.

What is increasingly being seen to be the case is that merely creating the right market conditions to achieve general environmental objectives, such as reducing carbon dioxide emissions, is not enough. Carbon taxes or emissions trading regimes are woefully insufficient on their own. Indeed, green activists in the EU (who have had most experience with the operation of carbon trading through the EU Emissions trading Scheme – EU ETS) have lost faith in carbon trading. It is seen as an institution that is being manipulated by the main energy industries and as a device that is incapable of assuring investment in capital-intensive eco-technologies because of uncertainty about the level of future energy prices (Clifton 2009).

While increases in energy taxation or prices are generally desirable on ecological grounds, it would be politically mistaken to assume that they can substitute for precisely targeted measures that can mobilise opinion for them on the basis of public identification with specific technologies. As is argued in Chapter 2, it is public identification with specific clean technologies that is necessary to ensure that ecological rationalities are given priority over economic rationalities. Planning instruments, regulations and price-setting incentive tools can fit in with this strategy, something that will not be achieved by either leaving technical choices up to mainstream industry or simply opposing technologies.

As the foregoing discussion indicates, what people call market forces or market conditions are an amalgam of what we need to call an array of rules introduced for planning, safety, environmental or engineering standard-setting reasons, as well as patterns of investment habits, company structures, consumer tastes and so on. To give incentives to renewable energy is not to move from free market competition to state intervention, but merely to shift the institutions that govern incentive structures from one type to another.

This set of lessons has been learned to an extent in the EU, and even in China. However, at the time of writing, in the USA, the movement in support of renewable energy is still clearly not strong enough to insist on the interventionist measures that are necessary to tackle long-term problems of energy sustainability. Part of the problem may be the USA's dominant free market political economy. Yet history has shown that when the USA has objectives of sufficiently high priority (for instance in wartime, or even in the 1970s when speed limits and energy efficiency standards were introduced for motor vehicles), then interventionist action will be taken. Certainly, in the UK, which has been associated with free market approaches to energy, there have been some interventionist measures in recent years in favour of renewable energy.

Further lines of research

Three lines of further research beckon. One, concerned with energy sustainability, is how 'identity EM' analysis can be deployed to understand how energy efficiency technologies have been, and can be, effectively promoted. Hitherto, in countries such as the UK and the USA energy efficiency programmes have been conducted largely on the basis of high-level bargaining between governments and various industries, the energy industries themselves about energy conservation subsidy programmes, and the building industry and appliance manufacturers about energy efficiency standards. However, as is discussed in Chapter 2, this 'mainstream' EM pattern of leaving technology choices largely up to the industry to be subsumed into otherwise 'business as usual' consumption patterns may simply be undermined by a general growth in demands for energy services. The example of renewable energy suggests that sustainable technologies, including energy efficiency technologies, will be most effectively delivered through campaigns, policy initiatives and policy instruments that specifically link the policies to the deployment of specific technologies. The key link is that a specific campaign or measure is associated with a specific technology, or at least a group of technologies, so that positive public identification with such technology can be mobilised. Of course, 'mainstream' EM modes of encouraging and mandating conventional industries to attain environmental objectives through technologies of their choice may help, but they are unlikely to be enough to cope with the global problems of resource depletion and pollution.

The second line of further research is into non-energy issues, where 'identity EM' may be relevant as a means of conceptualising ecological reform in other cases where there is a clear existing public identification of alternative technology. Organic agriculture is one that comes to mind. It shares some of the attributes of renewable energy, in that consumers positively identify with organic farming as a technology and that the industry is involved in a battle for market share with mainstream, chemically intensive agriculture.

The distinctions between identity EM and mainstream EM in this case, and also in the case of renewable energy, have a lot to do with the fact that renewable energy technologies such as wind power and solar power are seen, not just as another means of generating electricity, but as consumption choices in themselves. This is also the case with organic food, which is seen as a positive good in itself, not just another means of producing food. There is a hint of the 'Buddhist economics' of which Schumacher spoke. Rather than seeking to maximise consumption as an end in itself: 'since consumption is merely a means to human well-being, the aim should be obtain the maximum well-being with the minimum of consumption' (Schumacher 1973, 52). In this case, instead of maximising consumption, the development of eco-industries is minimising both pollution and resource depletion by using specific, preferred technologies.

However, the world has a long way to go before this approach is adopted with most consumption and most technologies. As was discussed in Chapter 6, in countries such as Spain and China, where there are rapidly expanding renewable programmes, there have also been rapidly expanding increases in electricity and energy demand. As was mentioned in Chapter 5, in the USA the rate of deployment of renewable energy will have to increase greatly before it can cancel out overall growth in electricity demand. On the other hand, examples such as Germany in the period 1990 to 2010 and Denmark in the 1970s to 2001 tentatively suggest there is a possibility of sustainable EM through its 'identity EM' variant. However, in these cases the relatively lower enthusiasm for new energy sources on the part of the major energy utilities (associated with relatively slower growth in energy demand) could only be countered by a strong social movement in support of renewable energy, some details of which are studied in another work (Toke 2011). The evidence is that a) energy sustainability is likely to be possible only in the context of 'identity EM', which is launched across a range of energy technologies on both the supply and demand side, and b) that social movements are required to be central parts of such strategies organised in support of a range of specific energy eco-technologies.

Interviews

Interviews appear in the order in which they were first cited in the book.

Renewable energy expert in Austin, 1 May 2009 (unattributable interview with leading renewable energy executive and activist in Texas).

Xiao-Ping Zhang, 24 February 2009. Dr Zhang is Reader in Energy Distribution Systems in the Department of Electronic, Electrical and Computer Engineering, University of Birmingham (UK).

Fraenkel, Peter, 17 July 2008. Dr Fraenkel is the Managing Director of Marine Current Turbines.

Taylor, Derek, 18 March 2009. Dr Derek Taylor is Principal of Altechnica, an energy-architectural design consultancy. He is also visiting Lecturer at the Energy and Environment Research Unit, Open University (UK).

Bowes-Lyon, Andy and Hairsine, Annie, 24 June 2009. Andy Bowes-Lyon and Annie Hairsine work for OTM Consulting, a company that has been given work to do desktop studies of vertical axis offshore wind turbine design.

Maegaard, Preben, 18 April 2009. From 1979 to 1984 Preben Maegaard was chairman of the Danish Renewable Energy Association, OVE.

Nagel, Joep, Tvind Folk High School 10 May 2009. Joep Nagal worked for many years as a teacher at Tvind, and took an active part in developing the Tvind Windmill.

Wellinger, Arthur, 6 January 2010. Arthur Wellinger has been involved in the biogas industry for several decades. He is now President of the European Biogas Association.

Green, Martin, 3 August 2009. Martin Green is Scientia Professor at the University of New South Wales.

Daniels, Lisa, 11 May 2009. Lisa Daniels is founder and currently Executive Director of Windustry, Minnesota.

Warren, Charles, 27 October 2009. Charles Warren was a member of the California State Assembly from 1963 to 1977.

Hamrin, Jan, 15 April 2009. Dr Hamrin founded and became Executive Director of the Independent Energy Producers' Association (IEP) in California.

Gipe, Paul, 6 May 2009. Paul Gipe organises 'Wind Works' and is a leading renewable energy author and campaigner in North America.

Sloan, Mike, 25 April 2009. Mike Sloan has been Executive Director of the Texas Wind Coalition.

Woolsey, Ed, 27 May 2009. Ed Woolsey is head of Green Prairie Energy, based in Prole, Iowa.

Lucas, Hugo (IDAE), 9 November 2009. Hugo Lucas is an Officer of the Institute for Energy Diversification and Saving, which is part of the Spanish Government.

Teske, Sven, Energy Campaigner for Greenpeace Germany, 14 September 2004.

Lipponen, Juho, 19 December 2007. Juho Lipponen is Head of Energy Policy and Generation Unit, Eurelectric.

Bischof, Ralf, BWE, 13 September 2004. Ralf Bischof was Managing Director of Bundesverband Windenergie, the German wind power trade association.

Holst, Henning, 18 July 1999. Henning Holst is an independent wind power consultant who helps communities set up wind power projects, especially in North Germany

Morris, Craig, 29 October 2009. Craig Morris is a US citizen who operates a business in Germany and is a renewable energy activist.

Puig, Josep, Eurosolar, 13 October 2009. Josep Puig was a founder member of Ecotechnia.

Cano, Alfonso, APPA, 9 December 2004. Alfonso Cano has been the Technical Advisor to the Spanish Renewable Energy Trade Association, the APPA.

Turmes, Claude, MEP, 27 February 2007. Claude Turmes is the MEP for Luxembourg and is Vice Chair of the Group of the Greens/European Free Alliance MEPs in the European Parliament.

Swift-Hook, Donald, 26 March 2009. Donald Swift-Hook was the former Head of the Applied Physics Laboratory at the Central Electricity Generating Board before retiring in 1987.

Lindley, David, 19 March 2009. David Lindley has been one of the leading organisers of wind power in the UK from the 1970s onwards. He founded National Wind Power, which later became Innogy, owned by RWE.

Leading civil servant dealing with energy policy, 12 November 1999.

Leading civil servant dealing with energy policy, 30 January 2008.

Former BWEA official, 3 October 2007 (unattributable interview).

O'Brien, Mike, Minister of State at the (UK) Department of Energy and Climate Change 2008–2010, 23 June 2010.

Dickson, Andrew, Wind Prospect, 5 August 2009. Andrew Dickson organises projects for Wind Prospect, a leading wind power developer in Australia.

Mohr, Tony, 29 June 2009. Tony Mohr is Climate Change Programme Manager for the Australian Conservation Foundation.

Milne, Christine, 21 May 2009. Christine Milne is a member of the Federal Senate of the Government of Australia. She is Deputy Leader of the Australian Green Party.

Executive of solar trading organisation, 14 August 2009 (unattributable interview).

Meyer, Sebastian, Research Director for the Beijing-based consultancy Azure International Technology & Development.

Shuang, Liu, 23 December 2009. Liu Shang is Energy and Climate Campaigner for Greenpeace, China.

Notes

1. Many would prefer to use the 'actor network theory' (ANT) formulation to describe how physical artefacts are involved in social systems. While I have no doubt that ANT is highly useful in many studies, I am also sympathetic to criticisms of ANT made by Collins and Yearley (1992, 317–22) about how different stories could be told about apparently the same network. In addition I sympathise with arguments made by Bloor (1999) that we should analyse the impact of physical artefacts through analysing what people believe about them.
2. A horizontal axis wind turbine is the type usually seen, while a vertical axis machine involves the blades moving around the tower rather as a merry-go-round revolves around a central vertical axis.
3. All statistics on wind power are taken from *Wind Power Monthly* except where otherwise indicated.

Bibliography

3E Consultants (2008) *A North Sea Electricity Grid Revolution*, Brussels, Belgium: Greenpeace.

Ackermann, T., Andersson, G. and Söder, L. (2001) Distributed Generation: A Definition, *Elec Power Syst Res*, 57, 195–204.

AEA Energy and Environment (2009) *Energy Statistics: Renewables*, London: Department of Energy and Climate Change, http://www.decc.gov.uk/en/content/cms/statistics/source/renewables/renewables.aspx

AEA Technology (2002) *Sea Wind East – How Offshore Wind In East Anglia Could Supply A Quarter of UK Electricity Needs*, London: Greenpeace.

Allchin, A. (1998) *N.F.S. Grundtvig: An Introduction to his Life and Work*, London: Darton, Longman and Todd, cited by M. Smith, *N.F.S. Grundtvig, Folk High Schools and Popular Education*, http://www.infed.org/thinkers/et-grund.htm, accessed June 2009.

Asmus, P. (2001) *Reaping the Wind*, Washington, DC: Island Press.

AWEA (2007) *U.S. Wind Energy Industry Applauds Federal Energy Regulatory Commission Ruling On New Transmission Policy*, http://www.awea.org/newsroom/releases/US_Wind_Energy_Industry_Applauds_FERC_Ruling_041907.html, accessed October 2009.

Barry, J. (1999) *Rethinking Green Politics*, London: Sage.

Barry, J. and Paterson, M. (2004) Globalisation, Ecological Modernisation and New Labour, *Political Studies*, 52, 767–84.

Beauregard-Tellier, F. (2007) *Electricity Production in China: Prospects and Global Effects*, Paris, France: Biblioteque du Parlement, http://www2.parl.gc.ca/Content/LOP/ResearchPublications/prb0704-e.pdf, accessed March 2010.

Bechberger, M. (2007) Why Renewable are not Enough: Spain's Discrepancy Between Renewable Growth and Energy (in)Efficiency, in L. Metz (ed.) *Green Power Markets: Support Schemes, Case Studies and Perspectives*, Brentwood UK: Multi-Science, pp. 201–26.

Beck, F., Hamrin, J., Brown, K., Sedano, R. and Singh, V. (2002) *Renewable Energy For California – Benefits, Status and Potential*, Washington, DC: Renewable Energy Policy Project.

Beck, U. (1992) *Risk Society*, London: Sage.

Beck, U. (1998) *Risk Society – Towards a New Modernity*, London: Sage.

Beeson, M. (2010 forthcoming) The Coming Of Environmental Authoritarianism, *Environmental Politics*.

Bell, D., Gray, T. and Haggett, C. (2005) Policy, Participation and the Social Gap in Wind Farm Siting Decisions, *Environmental Politics*, 14 (4), 460–77.

Bell, S. and Hindmoor, A. (2009) *Rethinking Governance*, Cambridge: Cambridge University Press.

Berry, T. and Jaccard, M. (2001) The Renewable Portfolio Standard: Design Considerations And An Implementation Survey, *Energy Policy*, 29, 263–77.

Bevir, M. and Rhodes, R. (2003) *Interpreting British Governance*, London: Routledge.

Bijker, W., (1995) *Of Bicycles, Bakelites and Bulbs: Toward a Theory of Sociotechnical Change*. Cambridge, MA: MIT Press.

Biogasgruppe (1990) *Biogas for Developing Countries*. http://nzdl.sadl.uleth. ca/cgi-bin/library?e=d-00000-00---off-0envl--00-0----0-10-0---0---0direct-10---4-------0-1l--11-en-50---20-about---00-0-1-00-0-0-11-1-0utfZz-8-00&a =d&cl=CL3.10&d=HASH01f0cfb7431f32f0413ea1e1.1, accessed September 2009.

Bird, L., Bolinger, M., Gagliano, T., Wiser, R., Brown, M. and Parsons, B. (2009) Policies and Market Factors Driving Wind Power Development in the United States, *Energy Policy*, 33, 1397–1407.

Blau, J. (2010) How Much PV Capacity Is Actually Installed in Germany?, *Renewable Energy World*, 21 January, http://www.renewableenergyworld.com/ rea/news/article/2010/01/fit-under-fire-are-german-pv-installation-numbers-purposely-underestimated?cmpid=WNL-Thursday-January21-2010, accessed January 2010.

Bloor, D. (1999) Anti Latour, *Study of the History of the Philosophy of Science*, 30 (1), 81–112.

Blühdorn, I (2000) Ecological Modernisation and Post-Ecologist Politics, in: F. Buttel, A. Mol and G. Spaargaren (eds) *Environment and Global Modernity*, London/Thousand Oaks/New Delhi: Sage, pp. 209–228.

Böhme, D. (ed.) (2009) *Development of Renewable Energy Sources in Germany in 2008*, Bonn: Federal Ministry for Environment, Nature Conservation and Nuclear Safety.

Bohn, C. and Lant, C. (2009) Welcoming the Wind? Determinants of wind power development among US states, *Prof. Geographer*, 61 (1), 87–100.

Borchers, A., Duke, J. and Parsons, G. (2007) Does Willingness To Pay For Green Energy Differ By Source?, *Energy Policy*, 35, 3327–34.

Boyle, G. (ed.) (2004) *Renewable Energy*, Oxford: OUP.

Brechin, S. (2003) Comparative Public Opinion and Knowledge on Global Climatic Change and the Kyoto Protocol: The U.S. versus the World? *International Journal of Sociology and Social Policy*, 23(10): 106–34.

Breukers, S. (2006) *Changing Institutional Landscapes for Implementing Wind Power*, Amsterdam: University of Amsterdam Press, pp. 177–221.

Broehl, J. (2008) A Subsidy Instead of a Tax Credit, *Wind Power Monthly*, 24 (11), 29.

Broehl, J. (2009) California Utilities Hide behind Solar Smokescreen, *Wind Power Monthly*, 25 (4).

Brower, M. (1992) *Cool Energy*, Cambridge, MA: The MIT Press.

Bunzel, J. (1983) *New Force on the Left – Tom Hayden and the Campaign Against Corporate America*, Palo Alto, CA: Stanford University/Hoover Institution Press.

Burgermeister, J. (2008) Biogas Flows Through Germany's Grid Big Time, *Renewable Energy World*, http://www.renewableenergyworld.com/rea/news/ article/2008/07/biogas-flows-through-germanys-grid-big-time-53075, accessed September 2009.

Business Insights (2010) *Green Energy In Germany*, London: Business Insights Ltd, www.globalbusinessinsights.com, accessed March 2010.

Business Wire (2006) *FPL Energy's California Wind Farms Have Positive Environmental and Economic Impact*, http://findarticles.com/p/articles/mi_ m0EIN/is_2006_Jan_30/ai_n16034571/, accessed October 2009.

Buttell, F. (2000) Ecological Modernization as Social Theory, *Geoforum*, 31, 57–65.

California Energy Commission (1997) *Critical Changes: California's Energy Plan*, Sacramento CEC, http://energyarchive.ca.gov/BR96/ENERGYPLAN.PDF, accessed September 2009.

California Energy Commission (2009) *California Energy Commission Energy Almanac*, http://energyalmanac.ca.gov/electricity/total_system_power.html, accessed September 2009.

Callon, M. (1986) Some Elements of Sociology of Translation: Domestication of the Scallops and the Fishermen of St Brieuc Bay, in J. Law (ed.) *Power Action and Belief – A New Sociology of Knowledge?*, London: Routledge and Kegan Paul, pp. 196–233.

Carter, N. (2001) *The Politics of the Environment – Ideas, Activism, Policy*, Cambridge: Cambridge University Press.

Christoff, P. (1996) Ecological Modernisation, Ecological Modernities, *Environmental Politics*, 5 (3), 476–500.

Clark, W. (1974) *Energy For Survival – The Alternative to Extinction*, New York: Anchor Books.

Clean Energy Council (2009) *Solar PV*, Melbourne: Clean Energy Council, http://www.cleanenergycouncil.org.au/cec/technologies/solarpv.html

Clean Techies (2009) (unsigned) *Desertec Is Taking Shape With 12 Companies Joining Consortium*, http://blog.cleantechies.com/2009/11/04/desertec-taking-shape-companies-joining-consortium/

Clifton, S.-J. (2009) *A Dangerous Obsession – The Evidence against Carbon Trading and for Real Solutions to Avoid a Climate Crunch – A Research Report*, London: Friends of the Earth, http://www.foe.co.uk/resource/reports/dangerous_obsession.pdf, accessed July 2010.

Coase, R. (1960) The Problem of Social Cost, *Journal of Law and Economics*, 3 (October), 1–44.

Cohen, B. (1983) Breeder Reactors: A Renewable Energy Source, *American Journal of Physics*, 51 (1), 75–6.

Collins, H. and Yearley, S. (1991) Epistemological Chickens, in A. Pickering (ed.), *Science As practice and Culture*, Chicago, IL: University of Chicago Press, pp. 301–26.

Connolly, K. (2009) German Blue Chip Firms Throw Weight Behind North African Solar Project, Guardian, 16 June, http://www.guardian.co.uk/environment/2009/jun/16/solar-power-europe-africa, accessed June 2009.

Cordsen, T. (2009) *Amendment of Feed-In Law Triggers Biogas Boom in Germany*, BORDA-Bremen, http://www.borda-net.org/modules/news/article.php?storyid=126, accessed December 2009.

CPUC (2006) Public Utilities Commission of the State of California *ENERGY DIVISION RESOLUTION E – 3980*, http://docs.cpuc.ca.gov/published/FINAL_RESOLUTION/55465.htm, accessed October 2009.

Czisch, G. (2006) *Low Cost but Totally Renewable Electricity Supply for a Huge Supply Area – a European/Trans-European Example*, http://transnational-renewables.org/Gregor_Czisch/projekte/LowCostEuropElSup_revised_for_AKE_2006.pdf, accessed March 2010.

De Groot, J. and Steg, L. (2008) Value Orientations to Explain Beliefs Related to Environmental Significant Behaviour: How to Measure Egoistic, Altruistic and Biospheric Value Orientations. *Environment and Behavior*, 40(3), 330–354.

Della Porta, D. and Diani, M. (1997) *Social Movements: An Introduction*, London: Blackwell.

Department of Energy and Climate Change (2009) *The UK Renewable Energy Strategy*, Cm 7686, London: Department of Energy and Climate Change, http://www.multi-science.co.uk/whycarbon.htm, accessed December 2009.

Department of Energy and Climate Change (2010) *Energy Consumption in the UK*, http://www.decc.gov.uk/en/content/cms/statistics/publications/ecuk/ecuk.aspx, accessed March 2010.

Department of Trade and Industry (1999a) *New and Renewable Energy – Prospects for the 21st Century*, London: DTI.

Department of Trade and Industry (1999b) *New and Renewable Energy – Prospects for the 21st Century – Analysis of the Responses to the Consultation Paper*, London: DTI.

Department of Trade and Industry (2000) *New and Renewable Energy – Prospects for the 21st Century – Preliminary Consultation*, The Renewables Obligation-Preliminary Consultation, London: DTI.

Derrida, J. (1976) *Of Grammatology*, Baltimore, MD: Johns Hopkins University Press.

Devine-Wright, P. (2007) Reconsidering Public Attitudes and Public Acceptance of Renewable Energy Technologies: A Critical Review, ESRC Working Paper 1.4, *Beyond Nimbyism: A Multidisciplinary Investigation of Public Engagement with Renewable Energy Technologies*, ESRC, http://www.sed.manchester.ac.uk/research/beyond_nimbyism/deliverables/bn_wp1_4.pdf

Devine-Wright, P. (2009) Rethinking Nimbyism: The Role of Place Attachment and Place Identity in Explaining Place Protective Action, *Journal of Community and Applied Social Psychology*, 19 (6), 426–41.

Dinica, V. (2008) Initiating a Sustained Diffusion of Wind Power: The Role of Public–Private Partnerships in Spain, *Energy Policy*, 36 (9), 3562–71.

Directorate General for Energy and Transport (2008) *Spain – Renewable Energy Fact Sheet*, Brussels: European Commission, http://www.energy.eu/renewables/factsheets/2008_res_sheet_spain_en.pdf

Dobson, A. (1995) *Green Political Thought*, London: Routledge.

Douglas, M. and Isherwood, B. (1979) *The World of Goods*, London: Allen Lane.

Douglas, M. and Wildavsky, A. (1982) *Risk and Culture*, Berkeley, CA: University of California Press.

Dryzek, S., Downes, D., Hunold, C. and Schlosberg, D., with Hernes, H.-K. (2003) *Green States and Social Movements: Environmentalism in the United States, United Kingdom, Germany and Norway*, Oxford: Oxford University Press.

DUKES (2010) *Digest of UK Energy Statistics*, London: Department of Energy and Climate Change, http://www.decc.gov.uk/en/content/cms/statistics/publications/dukes/dukes.aspx, accessed March 2010.

Eckersley, R. (1992) *Environmentalism and Political Theory*, London: UCL Press.

Econnect (2008) *Round 3 Offshore Wind Farm Connection Study*, London: The Crown Estate, http://www.thecrownestate.co.uk/round3_connection_study.pdf, accessed June 2009.

Economic Commission for Europe (1998) *Convention on Access to Information, Public Participation In Decision-Making And Access To Justice In Environmental Matters done at Aarhus, Denmark, on 25 June 1998,* http://www.unece.org/env/pp/documents/cep43e.pdf

ED (2009) The Cap and Trade Success Story, Environmental Defense Action Fund, 257 Park Avenue South, New York, NY10010. http://www.environmentaldefense.org/page.cfm?tagID=1085, accessed September 2009.

EDF (2009) *Renewable Energy Bill Passes Senate, Supporters Urge Quick Action by House,* http://www.edf.org/pressrelease.cfm?contentID=9756, accessed October 2009.

Energy Information Administration (2009) *US Electricity Data,* http://www.eia.doe.gov/fuelelectric.html, accessed October 2009.

Elliott, D. (1997) *Energy, Society and Environment,* London: Routledge.

ESAA (2003) *Submission by the Electricity Supply Association of Australia (ESAA) Limited to the Review of the Operation of the Renewable Energy (Electricity) Act 2000 (MRET Review),* Melbourne: ESAA. www.esaa.com.au, accessed November 2009.

ESAA (2007) Press Release: *ALP Interim Renewable Energy Target,* Melbourne: ESAA. www.esaa.com.au, accessed November 2009.

ESAA (2009) Submission to Senate Select Committee on Climate Policy, http://www.esaa.com.au/images/stories/policy_submissions/20090409senateclimatepolicy.pdf, accessed December 2009.

Eurobarometer 66-72 2007- 2010, EU Public Opinion, Commission of the European Communities, http://ec.europa.eu/public_opinion/index_en.htm, accessed July 2010.

Eyerman, R. and Jamison, A. (1991) *Social Movements: A Cognitive Approach,* University Park, PA: Penn State University Press.

Fischer, F. (2003) *Reframing Public Policy,* Oxford: Oxford University Press.

Fishkin, J. (1995) *Voice of the People – Public Opinion and Democracy,* New Haven, CT: Yale University Press.

Fouquet, D. (ed.) (2009) *Prices for Renewable Energies in Europe: Report 2009,* Brussels: European Renewable Energies Federation.

Fudge, C. and Rowe, J. (2001) Ecological Modernisation as a Framework for Sustainable Development: A Case Study in Sweden, *Environment and Planning A,* 33 (9), 1527–46.

Garud, R. and Karnoe, P. (2003) Bricolage versus Breakthrough: Distributed and Embedded Agency in Technology Entrepreneurship, *Research Policy,* 32, 277–300.

Geels, F. (2002) Technological Transitions as Evolutionary Reconfiguration Processes: A Multi-Level Perspective and a Case-Study, *Research Policy,* 31, 1257–74.

Gipe, P. (1995) *Wind Energy Comes of Age,* New York: John Wiley.

Gipe, P. (2003) Why I Oppose the Production Tax Credit, http://www.windworks.org/articles/lg_ProductionTaxCreditNo.html, accessed September 2009.

Global Watch Mission (2004) *Co-Operative Energy: Lessons from Denmark and Sweden,* London: DTI/Co-operatives UK.

Goldsmith, A.R., with additional help from Allaby, M., Davoll, J. and Lawrence, S. (1972) *Blueprint for Survival,* London: Penguin.

Gonzalez, J. and Freeman, S. (2009) Letter to Michael Peevey, Renewable Energy Accountability Project, http://www.reapinfo.org/ca-update1.html http://www.reapinfo.org/ca-update1.html

Green, M. (1993) Crystalline- and Polycrystalline- Silicon Solar Cells, in Johansson, T., Kelly, H., Reddy, A. and Williams, R. (eds) *Renewable Energy*, London: Earthscan, pp. 337–60.

Green, R. (2006) Electricity Liberalisation in Europe – How Competitive Will it Be?, *Energy Policy*, 34, 2532–41.

Greene, W. (1978) Who Owns 'AppropriateTechnology?' http://www.aliciapatterson.org/APF0101/Greene/Greene.html, accessed September 2009.

Greenpeace (2005) *Decentralising Power – An Energy Revolution for the 21st Century*, London: Greenpeace.

Habermas, J. (1971) *Toward a Rational Society*, London: Heinemann.

Habermas, J. (1973) *Legitimation Crisis*, Boston, MA: Beacon Press.

Habermas, J. (1987) *The Theory of Communicative Action*, Cambridge, MA: Polity.

Haggett, C. and Toke, D. (2006) Crossing the Great Divide – Using Multi-Method Analysis to Understand Opposition to Windfarms, *Public Administration*, 84 (1), 103–20.

Hajer, M. (1995) *The Politics of Environmental Discourse*, Oxford: Oxford University Press.

Hall, P. and Taylor, R. (1996) Political Science and the Three New Institutionalisms, *Political Studies*, XLIV, 936–57.

Hall, T. (1986) *Nuclear Politics*, London: Penguin

Hatch, (2007) The Politics of Climate Change in Germany: Domestic Sources of Environmental Foreign Policy, in Harris, P. (ed.) *Europe and Climate Change*, London: Edward Elgar, pp. 41–62.

Hau, E. (2006) *Wind Turbines – Fundamentals, Technologies, Application, Economics*, 2nd edn, Birkhäuser: Springer.

Hayden, T. (1980) *The American Future – New Visions Beyond Old Frontiers*, Boston MA: South End Press.

Helm, D. (2007) Labour's Third Energy White Paper, *Commentary*. www.dieter-helm.co.uk, accessed November 2008.

Hess, D. (2007) *Alternative Pathways in Science and Industry*, Cambridge, MA: MIT Press.

Heymann, M. (1998) The Signs of Hubris: The Shaping of Wind technology Styles in Germany, Denmark, and the United States, 1940–1990, *Technology and Culture*, 39 (4), 641–70.

Hodgson, M. (1998) The Approach of Institutional Economics, *Journal of Economic Literature*, 36, 166–92.

Holdren, J. and Herrera, P. (1971) *Energy*, San Francisco: Sierra Club.

Howarth, D. (2000) *Discourse*, Buckingham: Open University Press.

Huber, J. (1991) Ecologische modernisering; weg van schaartse, soberheid en bureaucratie?, in A. Mol, G. Spaargaren and A. Klapwijk (eds) *Technologie en milieubeheer: Tussen sanering en ecologische moderisering*, Den Haag: SDU.

Huber, J. (2004) *New Technologies and Environmental Innovation*, Cheltenham, UK: Edward Elgar.

Hughes, T. (1983) *Networks of Power: Electrification in Western Society, 1880–1930*, Baltimore: John Hopkins University Press.

Hughes, T. (1987) The Evolution of Large Technical Systems, in W. Bijker, T. Hughes and T. Pinch (eds), *The Social Construction of Technological Systems*, Cambridge, MA: The MIT Press.

Hurlbut, D. (2008) A Look Behind the Texas Renewable Portfolio Standard: A Case Study, *Natural Resources Journal*, 48 (1), 129–62.

Hvelplund, F. (2005) Renewable Energy: Political Prices or Political Quantities, in V. Lauber (ed.) *Switching to Renewable Power*, London; Earthscan, pp. 228–45.

Instituto para la Diversificación y Ahorro de la Energía (IDAE) (2005) *Plan de energies renovables en espana 2005-2010*, Madrid: Ministerio de Industria Turismo y Comercio.

Irwin, A. (1995) *Citizen Science*, London: Routledge.

Jacobsson, S. and Bergek, K. (2003) Transforming the Energy Sector: The Evolution of Technical Systems in Renewable Energy Technology, *Industrial and Corporate Change*, 13 (5), 815–27.

Jacobsson, S. and Lauber, V. (2005) Germany: From a Modest Feed-in Law to a Framework for Transition, in V. Lauber (ed.) *Switching to Renewable Power*, Earthscan, pp. 122–58.

Jacobsson, S., Sandén, B, and Bångens, L. (2004) Transforming the Energy System—the Evolution of the German Technological System for Solar Cells. *Technology Analysis & Strategic Management*, 16 (1), pp. 3–9.

Jamison, A. (2001) *The Making of Green Knowledge. Environmental Politics and Cultural Transformation*, Cambridge: Cambridge University Press.

Jamison, A. (2006) Social Movements and Science: Cultural Appropriations of Cognitive Praxis, *Science as Culture*, 15 (1), 45–59.

Jamison, A., Eyerman, R., Cramer, C. and Læssoe, J. (1990) *The Making of the New Environmental Consciousness: A Comparative Study of the Environmental Movements in Sweden, Denmark and the Netherlands*, Edinburgh: Edinburgh University Press.

Janicke, M. (2008) Ecological Modernisation: New Perspectives, *Journal of Cleaner Production*, 16, 557–65.

Jasanoff, S. (1990) *The Fifth Branch: Science Advisors as Policymakers*, Cambridge, MA: Harvard University Press.

Jasper, J. (1990) *Nuclear Politics*, Princeton, NJ: Princeton University Press.

Johnson, K. (2009) FERC Chairman: We Don't Need No Stinkin' Nukes, *Wall Street Journal*, online edition, 23 April. http://blogs.wsj.com/environmentalcapital/2009/04/23/ferc-chairman-we-dont-need-no-stinkin-nukes/, accessed May 2009.

Jungk, R. (1979) *The Nuclear State*, London: John Calder.

Karger, C. and Hennings, W. (2009) Sustainability Evaluation of Decentralized Electricity Generation, *Renewable and Sustainable Energy Reviews*, 13 (3), 583–93.

Karnoe, P. (1990) Technological Innovation and Industrial Organisation in the Danish Wind Industry, *Entrepreneurship and Regional Development*, 2, 105–23.

Kelly, G. (2007) Renewable Energy Strategies in England, Australia and New Zealand, *Geoforum*, 38, 326–38.

Kemp, R., Schot, J.W. and Hoogma, R. (1998) Regime Shifts to Sustainability through Processes of Niche Formation: The Approach of Strategic Niche Management, *Technology Analysis and Strategic Management*, 10, 175–96.

Kingdon, J. (1984) *Agendas, Alternatives and Public Policies,* New York: HarperCollins.

Kirschner, D., Barkovich, B., Treleven, K. and Walther, R. (1997) A Cost-Effective Renewables Policy Can Advance the Transition to Competition, *The Electricity Journal,* 10(1), 54–61.

Komanoff, C. (1981) *Nuclear Power Plant Cost Escalation,* New York: Charles Komanoff Energy Associates.

Kuhn, T. (1970) *The Structure of Scientific Revolutions,* Chicago, IL: University of Chicago Press.

Laclau, E. (1980) *New Reflections on the Revolution of Our Times,* London: Verso.

Lancaster, T. (1989) *Policy Stability and Change – Energy in Spain's Transition,* London: Pennsylvania State University.

Langhelle, O. (2000) Why Ecological Modernisation and Sustainable Development Should Not Be Conflated, *Journal of Environmental Policy and Planning,* 2, 303–22.

Langniss, O. and Wiser, R. (2003) The Renewables Portfolio Standard in Texas: An Early Assessment, *Energy Policy,* 31, 527–35.

Lauber, V. (2007) Anglo Saxon and German Approaches to Neo-Liberalism and Environmental Policy: The Case of Financing Renewable Energy, *Geoforum,* 38(4), 677–87.

Lauber, V. and Metz, L. (2004) Three Decades of Renewable Electricity Policies in Germany, *Energy & Environment,* 15 (4), 599–623.

Leach, G., Lewis, C., Romig, A., van Buren, A. and Foley, G. (1979) *Low Energy Strategy for the UK,* London: International Institute for Environment and Development.

Li, J., Shi, Jingli, X. Hongwen, Song, Yanqin and Shi, Pengfei (2006) *A Study on the Pricing Policy of Wind Power in China,* Beijing: Chinese Renewable Energy Industries Association (CREIA), Greenpeace, Global Wind Energy Council.

Light, A. (2000) What is an Ecological Identity?, *Environmental Politics,* 9 (4), 59–81.

Lovell, H., Bulkeley, H. and Owens, S. (2009) Converging Agendas? Energy and Climate Change Policies in the UK, *Environment and Planning C: Government and Policy,* 27 (1), 90–109.

Lovins, A. (1977) *Soft Energy Paths – Towards a Durable Peace,* London: Pelican.

McDonald, S., Alevizou, P., Oates, C., Hwang, K. and Young, W. (2007) Decoding Governance: A Study of Purchasing Processes for Sustainable Technologies, in Murphy, J (ed.) *Governing for Sustainability,* London: Earthscan.

McGovern, M. (2004) Spain Gets it Together, *Wind Power Monthly,* 20 (10), 6.

MacKay, D. (2009) *Sustainable Energy – Without the Hot Air,* Cambridge: UIT Cambridge.

Mah, D.-N. (2010) Collaborative Governance for Sustainable Development: Wind Resource Assessment in Xinjiang and Guangdong Provinces, China, *Sustainable Development,* forthcoming.

Mao, Y., Sheng, H. and Yan, F. (2008) *The True Cost of Coal,* Beijing: Greenpeace, China Sustainable Energy Programme, WWF, http://www.greenpeace.org/raw/content/china/en/press/reports/the-true-cost-of-coal.pdf, accessed January 2010.

March, J. and Olsen, J. (1989) *Rediscovering Institutions: The Organizational Basis of Politics*, New York: Free Press.

Markovits, A. and Gorski, P. (1993) *The German Left—Red Green and Beyond*, Cambridge: Polity Press.

Marsh, D. and Rhodes, R.A.W. (eds) (1992) *Policy Networks in British Government*, Oxford: Oxford University Press.

Melloan, G. (1987) Californians Will Pay Dearly for Purpa Power, *Wall Street Journal*, 31 March 1987, p. 37, http://www.fortfreedom.org/p11.htm, accessed September 2009.

Melucci, A. (1995) The Process of Collective Identity, in H. Johnston and B. Klandermans (eds), *Social Movements and Culture*, London: UCL.

Melucci, A. (1996) *Challenging Codes: Collective Action in the Information Age*, Cambridge: Cambridge University Press.

Mendonca, M., Jacobs, D. and Sovacool, B. (2010) *Powering the Green Economy – The Feed-In Tariff Handbook*, London: Earthscan.

Milanez, B. and Bührs, T. (2008) Ecological Modernisation beyond Western Europe: The Case of Brazil, *Environmental Politics*, 17 (5), 784–803.

Ministry of Fuel and Power (1955) *A Programme of Nuclear Power* CMD 9389, London: Her Majesty's Stationery Office.

Mitchell, C. (2000) The England and Wales Non-Fossil Fuel Obligation, *Annual Review of Energy and Environment*, 25, 285–312.

Mitchell, C. (2008) *The Political Economy of Sustainable Energy*, London: Palgrave.

Mol, A. (1995) *The Refinement of Production – Ecological Modernisation Theory and the Chemical Industry*, Utrecht: Van Arkel.

Mol, A. (1996) Ecological Modernisation and Institutional Reflexivity: Environmental Reform in the Late Modern Age, *Environmental Politics*, 5 (2), 302–23.

Mol, A. (2001) *Globalization and Environmental Reform: The Ecological Modernisation of the Global Economy*, Cambridge, MA: MIT Press.

Mol, A. (2006) Environment and Modernity in Transitional China: Frontiers of Ecological Modernization, *Development and Change*, 37 (1), 29–56.

Mol, A. and Spagaaren, G. (2000) Ecological Modernisation Theory in Debate: A Review, *Environmental Politics*, 9(1), 17–49.

Mouffe, C. (1995) Post Marxism: Democracy and Identity, *Environment and Planning D: Society and Space*, 13 (3), 259–65.

Murphy, J. (2000) Editorial Ecological Modernisation, *Geoforum*, 31, 1–8.

Murphy, J. and Gouldson, A. (2000) Environmental Policy and Industrial Innovation: Integrating Environment and Economy through Ecological Modernisation, *Geoforum*, 31, 33–44.

Ngan, H. (2010) Electricity Regulation and Electricity Market Reforms in China, *Energy Policy*, 38, 2142–8.

Nielsen, K. (2001) *Tilting at Windmills: On Actor-Worlds, Socio-Logics and Techno-Economic Networks of Wind Power in Denmark 1974–1999*, PhD thesis, University of Aarhus.

North Carolina Solar Energy Association (2009) *Bay Area Solar Installation Report*, http://www.norcalsolar.org/index.php?option=com_content&task=view&id=107&Itemid=181, accessed October 2009.

Norton, G. (1994) *Towards Unity Among Environmentalists*, Oxford: Oxford University Press.

NREL (2010) *Department of Energy Releases New Estimates of Nation's Wind Energy Potential*, 26 February 2010, http://www.nrel.gov/wind/news/2010/816.html, accessed March 2010.

O'Brien, M. (2009) Speech to *UK Ports: Opportunities from the Offshore Wind Industry*, Conference held at BERR Conference Centre, 1 Victoria St, 30 March.

Offe, C. (1985) New Social Movements: Changing Boundaries of the Political, *Social Research*, 52, 817–68.

Office of Appropriate Technology (1980) *Conservation and Solar Energy Programs of the Department of Energy: A Critique by the Office of Appropriate Technology, State of California*, June 1980, Washington, DC: Department of Energy, http://www.fas.org/ota/reports/8006.pdf, accessed September 2009.

Office of the Governor Rick Perry (2008) *Texas Needs To Lead The Charge On Energy Independence*, Press Release, 9 October, http://governor.state.tx.us/news/press-release/11395/#, accessed October 2009.

Office of the Renewable Energy Regulator (2009) *About RET*, Canberra: Australian Government, http://www.orer.gov.au/

Olesen, G. (1998) *Large Scale Implementation of Renewable and Sustainable Energy*, Hjortshoj, Denmark: OVE/INFORSE-Europe.

Patterson, W. (1976) *Nuclear Power*, London: Pelican.

Perlin, J. (1999) *From Space to Earth – The Story of Solar Electricity*, Ann Arbor, MI: aatec Press.

Peters, G. (2002) *Governance: A Garbage Can Perspective*, Vienna: Institute for Advanced Studies, http://www.ihs.ac.at/publications/pol/wp_84.pdf, accessed June 2009.

Pierson, P. and Skocpol, T. (2002) Historical Institutionalism in Contemporary Political Science, in I. Katznelson and H.V. Milner (eds), *Political Science: State of the Discipline*, New York: W.W. Norton, pp. 693–721.

Pope, C. (2007) *Taking the Initiative*, San Francisco: Sierra Club, http://sierraclub.typepad.com/carlpope/2007/07/the-capital-of.html, accessed October 2009.

Price, T. (2005) James Blyth – Britain's First Modern Wind Power Pioneer, *Wind Engineering*, 29 (3), 191–200.

Puig, J. (2009) Renewable Regions: Life after Fossil Fuels in Spain, in P. Droege (ed.), *100% Renewable – Energy Autonomy in Action*, London: Earthscan, pp. 187–204.

Pursell, C. (2001) *American Technology*, Malden, MA: Blackwell.

Putnam, P. (1948) *Power from the Wind*, New York: Van Nostrand Reinhold.

Qi, Wu (2010) Grid Quotas Aim to Connect 9 GW of Stuck Chinese Wind, and Chinese Sector Battered by Ministerial Sandstorm, *Wind Power Monthly*, 26 (6), 17, 84–6.

Quiang Wang (2010) Effective Policies for Renewable Energy – The Example of China's Wind Power – Lessons for China's Photovoltaic Power, *Renewable and Sustainable Energy Reviews*, forthcoming.

Rader, N. (2000) The Hazards of Implementing Renewables Portfolio Standards, *Energy & Environment*, 11 (4), 391–405.

Rader, N. and Norgaard, R. (1996) Efficiency and Sustainability in Restructured Electricity Markets: The Renewable Portfolio Standard, *The Electricity Journal*, 9 (6), 37–49.

Raven, R. (2004) *Strategic Niche Management for Biomass*, Eindhoven: Eindhoven Centre for Innovation Studies.

Renn, O., Webler, T. and Wiedemann, P. (1995) *Fairness and Competence in Citizen Participation*, Boston, MA: Kluwer.

Reuters (2009) *Research.and.Markets*, http://www.reuters.com/article/pressRelease/idUS131894+23-Feb-2009+BW20090223, accessed September 2009.

Rhodes, R. (1997) *Understanding Governance*, Milton Keynes: Open University Press.

Righter, R. (1996) *Wind Energy in America – A History*, Norman: University of Oklahoma Press.

Rosen, D. and Houser, T. (2007) *China Energy – A Guide for the Perplexed*, Peterson Institute for International Economics, http://www.iie.com/publications/papers/rosen0507.pdf, accessed January 2010.

Ross, D. (1995) *Power from the Waves*, Oxford: Oxford University Press.

Roe, D. (1984) *Virgins and Dynamos*, New York: Random House.

Rudig, W. (1990) *Anti-Nuclear Movements: A World Survey of Opposition to Nuclear Energy*, Harlow, Essex: Longman.

Ryland, E. (2010) Danish Wind Power Policy: Domestic and International Forces, *Environmental Politics*, 19 (1), 80–5.

Sachs, Wolfgang (ed.) *Global Ecology: A New Arena of Political Conflict*, London: Zed Books.

Salter, S. (2000) *Memorandum to the House of Commons Select Committee on Science and Technology*, Seventh Report of Session 2000–2001, on Wave and Tidal Energy, HC 291.

Schaefer, B. (2006) *Cultural Influences on Renewable Energy Acceptance and Tools for the Development of Communication Strategies to Promote Acceptance among Key Actor Groups*, Barcelona: Ecoinstitut Barcelona, http://www.createacceptance.net/fileadmin/create-acceptance/user/docs/CASE_16.pdf, accessed January 2010.

Scheer, H. (2007) *Energy Autonomy*, London: Earthscan.

Scheer, H. (2008) Hermann Scheer's Arguments against the Supergrid 2.12.08, email to Bianca Jagger (President of World Future Council) from Nina Alsen, 2 December 2008.

Schmidt, V. (2008) Discursive Institutionalism: The Explanatory Power of Ideas and Discourse, *Annual Review of Political Science*, 11(June).

Schumacher, F. (1972) The Work of the Intermediate Technology Development Group in Africa, *International Labour Review*, 106 (1), 75–92.

Schumacher, F. (1973) *Small Is Beautiful – A Study of Economics as if People Mattered*, London: Blond and Briggs.

Schumpeter, J. (1975) *Capitalism, Socialism and Democracy*, New York: Harper, first published 1942, pp. 82–5, http://transcriptions.english.ucsb.edu/archive/courses/liu/english25/materials/schumpeter.html, accessed September 2009.

Scott, A. (1990) *Ideology and Social Movement*, London: Unwin Hyman.

Scott, R. (1995) *Institutions and Organisation*, Thousand Oaks, CA: Sage.

SenterNovem (2005) *European Concerted Action for Offshore Wind Energy Deployment* (COD)- Work Package 8 – Grid Issues, Brussels: Commission of the European Communities, DG 17, http://www.offshorewindenergy.org/cod/CODReport_Grid.pdf, accessed June 2009.

Shackley, S., McLachlan, C. and Gough, C. (2005) The Public Perception of Carbon Capture and Storage in the UK, *Climate Policy*, 4, 377–98.

Shiva, V. (1989) *Staying Alive: Women, Ecology and Development*, London: Zed Books.

Shove, E. (2003) *Comfort Cleanliness + Convenience – The Social Organisation of Normality*, London: Sage.

Shove, E. (2006) Efficiency and Consumption: Technology and Practice, in T. Jackson (ed.), *Sustainable Consumption*, London: Earthscan, pp. 293–304.

Sierra Club California (2002) *Support Clean Energy in California and Fight Global Warming!*, http://www.sierraclub.org/ca/energy/takeaction.asp, accessed October 2009.

Sinovel (2010) *Company Overview*, Beijing: Sinovel World Group, http://www.sinovel.com/Companyoverview2.html, accessed March 2010.

Smith, A. (2004) Alternative Technology Niches and Sustainable Development, *Innovation: Management, Policy & Practice*, 6, 220–35.

Smith, A. (2007) Translating Sustainabilities between Green Niches and Socio-Technical Regimes, *Technology Analysis & Strategic Management*, 19 (4), 427–50.

Smith, A., Stirling, A. and Berkhout, F. (2005) The Governance of Sustainable Socio-Technical Transitions, *Research Policy*, 34, 1491–1510.

Sovacool, B. (2008) *The Dirty Energy Dilemma – What's Blocking Clean Energy in the USA*, Westport, CT: Praeger.

Spaargaren, G. (2000) Ecological Modernisation Theory and Changing Discourses on Environment and Modernity, in G. Spaargaren, A. Mol and F. Buttel (eds), *Environment and Modernity*, London: Sage, pp. 41–72.

Star, S. and Griesemer, J. (1989) Institutional Ecology, 'Translations' and Boundary Objects: Amateurs and Professionals in Berkeley's Museum of Vertebrate Zoology, 1907–39, *Social Studies of Science*, 19 (4), 387–420.

Starrs, T. (1988) Legislative Incentives and Energy Technologies: Government's Role in the Development of the California Wind Energy Industry, *Ecology Law Quarterly*, 15, 103–58.

Stenzel, T. and Frenzel, A. (2008) Regulating Technological Change – The Strategic Reactions of Utility Companies towards Subsidy Policies in the German, Spanish and UK Electricity Markets, *Energy Policy*, 36 (7), 2645–57.

Streeck, W. and Thelen, A. (2005) *Beyond Continuity: Institutional Change in Advanced Political Economies*, Oxford: Oxford University Press.

Szarka, J. (2007) *Wind Power in Europe – Politics Business and Society*, London: Palgrave.

Tech-Wise (2002) ReviewofCurrentEUandMSElectricityPolicyandRegulation – Denmark, http://www.ecn.nl/fileadmin/ecn/units/bs/SUSTELNET/wp2_denmark-sc.pdf, accessed June 2009.

Toke, D. (2002a) Ecological Modernisation and GM Food, *Environmental Politics*, 11 (3), 145–63.

Toke, D. (2002b) Wind Power in UK and Denmark: Can Rational Choice Theory Help Explain Differences, *Environmental Politics*, 11 (4), 83–100.

Toke, D. (2005) Explaining Wind Power Planning Outcomes, Some Findings from a Study in England and Wales, *Energy Policy*, 33 (12), 1527–39. Toke, D. (2010) Politics by Heuristics – Policy Networks with a Focus on Actor Resources, as Illustrated by the Case of Renewable Energy Policy Under New Labour, *Public Administration*, forthcoming.

Toke, D. (2011) Ecological Modernisation, Social Movements and Renewable Energy, *Environmental Politics*, 20 (1), forthcoming.

Toke, D. and Strachan, P. (2006) Wind Power and Ecological Modernisation, *European Environment*, 16, 155–166.

Toke, D., Breukers, S. and Wolsink, M. (2008) Wind Power Deployment Outcomes: How Can We Account for the Differences?, *Renewable and Sustainable Energy Reviews*, 12, 1129–47.

Touraine, A., Hegedus, Z., Dubet, F. and Wieviorka, M. (1983) *Anti-Nuclear Protest: The Opposition to Nuclear Energy in France*, Cambridge: Cambridge University Press.

TPUC (2009) Public Utilities Commission of the State of Texas, *New Electric Generating Plants in Texas – New Generating Plants in Texas since 1995*, http://www.puc.state.tx.us/electric/maps/gentable.pdf, accessed October 2009.

TREI (2009) *Texas Energy Independence Week*, http://www.treia.org/mc/page.do?sitePageId=84104, accessed October 2009.

UNECE (1998) *Convention on Access to Information, Public Participation in Decision-Making and Access to Justice in Environmental Matters Done at Aarhus, Denmark, on 25 June 1998*, Geneva, Switzerland: United Nations, http://www.unece.org/env/pp/documents/cep43e.pdf, accessed March 2010.

Unruh, G. (2000) Understanding Carbon Lock-In, *Energy Policy*, 28, 817–30.

USDOE (2003) *Wind Pioneer Interview: Jim Dehlsen, Clipper Windpower*, USDOE, Washington, DC: Energy Efficiency and Renewable Energy Program, http://www.windpoweringamerica.gov/filter_detail.asp?itemid=683, accessed March 2010.

US EIA – Energy Information Administration (2010) *Electric Power Industry 2008: Year in Review*, Table ES1. Summary Statistics for the United States, 1997 through 2008, http://www.eia.doe.gov/cneaf/electricity/epa/epaxlfilees1.pdf, accessed March 2010.

Van de Poel, I. (2000) On the Role of Outsiders in Technical Development, *Technology Analysis & Strategic Management*, 12 (1), 383–97.

Van der Horst, D. (2007) Exploring the Relevance of Location and the Politics of Voiced Opinions in Renewable Energy Siting Controversies, *Energy Policy*, 35, 2705–14.

van Est, R. (1999) *Winds of Change – A Comparative Study of Politics of Wind Energy Innovation in California and Denmark*, Utrecht: International Books.

Veblen, T. (1970) *THE theory of the Leisure Class: An Economic Study of Institutions*, London: Allen and Unwin.

Vote Solar (2009) *Press Release: CPUC Proposes Bold New Solar Energy Program for California*, http://www.votesolar.org/press/CPUCFiT.html, accessed October 2009.

Watt, R. (1998) Towards a Synthesised Network Approach: An Analysis of UK Nuclear and Renewable (Wave) Energy Programmes 1939–1985, a thesis submitted to the Faculty of Arts of the University of Birmingham for the degree of Doctor Of Philosophy.

Weale, A. (1992) *The New Politics of Pollution*, Manchester: Manchester University Press.

Weart, S. (1988) *Nuclear Fear – A History of Images*, Cambridge, MA: Harvard University Press.

Wellock, T. (1998) *Critical Masses – Opposition to Nuclear Power in California, 1958–78*, Madison, WI: University of Wisconsin Press.

Wiser, R. and Bolinger, M. (2008) *Annual Report on U.S. Wind Power Installation, Cost, and Performance Trends: 2007*, Washington, DC: Department of Energy.

Wiser, R. and Langniss, O. (2003) The Renewables Portfolio Standard in Texas: An Early Assessment, *Energy Policy*, 31, 527–35.

Wiser, R., Bolinger, M. and Raitt, H. (2005) Does It Have To Be This Hard? Implementing the Nation's Most Complex Renewables Portfolio Standard, *The Electricity Journal*, 18 (8), 55–7..

Wiser, R., Pickle, S. and Goldman, C. (1998) Renewable Energy Policy and Electricity Restructuring: A California Case Study, *Energy Policy*, 26 (6), 465–75.

Wolfe, P. (2008) The Implications of an Increasingly Decentralised Energy System, *Energy Policy*, 36, 4509–13.

Wolsink, M. (2000) Wind Power and the NIMBY-Myth: Institutional Capacity and the Limited Signifcance of Public Support, *Renewable Energy*, 21, 49–64.

Wolsink, M. (2007) Planning of Renewables Schemes: Deliberative and Fair Decision-Making on Landscape Issues Instead of Reproachful Accusations of Non-Cooperation, *Energy Policy*, 35, 2692–704.

Woo, C.-K. (2001) What Went Wrong in California's Electricity Market?, *Energy*, 26, 747–58.

World Bank (2009) *Different Forms of Decentralisation*, http://www.ciesin.org/decentralization/English/General/Different_forms.html, accessed June 2009.

World Nuclear Association (2010) *Australia's Electricity*, http://www.world-nuclear.org/info/inf64.html, accessed March 2010.

Wu Qi (2009) China Opens Market to Foreign Manufacturers, *Wind Power Monthly*, 25 (12), 27–8.

Zhang, X., Ruoshui, W., Molin and Martinot, E. (2010) A Study of the Role Played by Renewable Energies in China's Sustainable Energy Supply, *Energy*, forthcoming.

Zografos, C. and Martinez-Alier, J. (2009) The Politics of Landscape Value: A Case Study of Wind Farm Conflict in Rural Catalonia, *Environment and Planning A*, 41 (7), 1726–44.

Index